Information Flow and Knowledge Sharing

Capturing Intelligence

Series Editors

Frank van Harmelen, Vrije Universiteit Amsterdam
Didier Dubois, Université Paul Sabatier Toulouse
John Taylor, King's College London
Sebastian Thrun, Stanford University

ELSEVIER

Amsterdam Boston Heidelberg London New York Oxford
Paris San Diego San Francisco Singapore Sydney Tokyo

Information Flow and Knowledge Sharing

Flávio Soares Corrêa da Silva

Jaume Agustí-Cullell

ELSEVIER

Amsterdam Boston Heidelberg London New York Oxford
Paris San Diego San Francisco Singapore Sydney Tokyo

Elsevier
Radarweg 29, P.O. Box 211, 1000 AE Amsterdam, the Netherlands
The Boulevard, Langford Lane, Kidlington, Oxford OX5 1GB, UK

For information on all Elsevier Academic Press publications
visit our Web site at www.books.elsevier.com

ISBN: 978-0-444-52935-0
ISSN: 1574-9576

Transferred to Digital Printing 2011

Contents

Foreword

The stated aim of the book series "Capturing Intelligence" is to publish books on research from all disciplines dealing with and affecting the issue of understanding and reproducing intelligence in artificial systems. Of course, much of the work done in the past decades in this area has been of a highly technical nature, varying from hardware design for robots, software design for intelligent agents, and formal logic for reasoning.

It is therefore very refreshing to see the book written by Jaume Agusti-Cullell and Flavio Soares Correa da Silva, entitled "Information Flow and Knowledge Sharing". This is a courageous book indeed. It is not afraid to tackle the Big Issues: notions such as information, knowledge, information system, information flow, collaborative problem solving, and ontological reasoning. All of these notions are crucial to our understanding of intelligence and our building of intelligent artificial systems, but all too often, these Big Issues are hidden behind the curtains while the technical topics take center stage. But these big foundational issues do indeed influence how work is done on the technical stage. In our everyday technical and scientific practice we use words like "knowledge system" and "information system", or "knowledge base", and "data base" without asking ourselves if we actually understand words such as "knowledge", "data", and "information". This book aims at no less then a conceptualization of these two central terms "information" and "knowledge", and to clarify the distinctions between these two concepts.

The book is not so much technical as it is philosophical, which is perhaps somewhat non-standard in this series. And the authors have chosen a correspondingly non-standard format. Their philosophical and conceptual analyses are punctuated by interludes, short stories that serve both to entertain and to tickle the reader. They reflect on material that has been presented before, and prepare the ground for the questions that are to follow. AI has a rich history of philosophical books that have chosen a non-standard structure and narrative (with Douglas Hofstadter as the champion of this genre). It is nice to see that the authors have succeeded into combining a non-standard approach to deep questions with a non-standard format, resulting in a highly interesting volume.

Frank van Harmelen, Series Editor

Acknowledgments

The authors benefited, during the preparation of this book, from financial support granted by the Spanish CSIC and the Brazilian CNPq and FAPESP. We thank the support provided by the Series Editor, Prof. Frank van Harmelen, as well as by our Editor at Elsevier, Mary James.

Flávio thanks his wife, Renata, and his daughter, Maria Clara, for the patience and support during the preparation of the book. They were, after all, the reason for him to keep writing and wishing to have things accomplished.

Jaume thanks his family, who were his unending source of inspiration for carrying out work that, after all, is done on their behalf.

List of Figures

Prologue

Beep, beep, beep...

Again.

Six o'clock.

Again.

Time to get started...

Again.

My wife and daughter are still asleep. They will probably sleep till nine or so. As for me, it is time to wake up. As usual, I am overloaded with work, and there are a few administrative stuff I would like to have done before my lecture starts.

Luckily, I live very close to the university. This means that I can leave home twenty minutes before eight, and I will certainly be starting my lecture punctually at eight.

I am teaching Introduction to Computer Science to first-year students this year. It is curious to see how this course is given little attention in our department. I vaguely recall reading about some university (in the Netherlands, I think) that reserved the first year courses to the most experienced professors, which is quite the inverse of what we do in our department. My stubbornness in teaching freshers is put by my colleagues together with my interest in philosophy, ancient music, computer games and Tai Chi. I have heard along the years comments ranging from discreet ironies to blatant criticisms about my non-professional approach to academic career, and as time went by I somehow got used to them. Quite often these comments appeared in discussions about the relative importance of teaching versus research, about the relative importance and time to be allotted to PhD students versus undergraduate students, or about who should be working for who when a lecturer supervises a PhD student (should the lecturer consider him/herself at the service of the student, or should him/her assume that the student would be at the service of his/her research pursuits and goals?).

My interest in teaching the freshers is generally taken with appreciation, though, for very practical reasons. Every time I take a first year course, my colleagues are relieved from one more first year course to teach. I am happy, everybody is happy.

My reasons to ask for such courses relate to these youngsters still not having a clear idea of what the university is about, nor being so much compelled by "practical concerns" such as how to get trained to find a job that pays well. Hence, they are still interested in discussing what the university may be about, and perhaps what it should be about. Indeed, trying to understand what the university is about seems to be a topic for discussion among freshers and close to retiring professors. These two end points of the academic career seem to share, if nothing else, shortened interest about "practical matters", which curiously leads them to discussing those matters of greatest practical reach in the long term.

My first year students still do not have a clear idea of what science is, or what technology is, or how they are related to each other. And they are indeed interested in learning about these issues. They are not tied to concepts they may have learned to be the good ones to show off in a conversation. They still consider these issues of interest, some of them from a very pragmatic point of view (since they have already been taught that this may be important for their professional preparation), but others just for the sake of knowing them. Hence, they are still interested in discussing these issues.

It is certainly not my task to lead such discussions with these students. We have work to do. I have to teach them the rudiments of programming in the next four months. But I can use without causing any disruption to the course some five to ten minutes in a lecture every now and again to let some discussion flow about these topics. I certainly have no answers to give them, but I believe it can be healthy for them to see their gray-haired teacher still looking for answers after all these years. If for nothing else, at least they may consider that these issues may be not simple at all, and somehow interesting enough to keep someone like their teacher interested in them for so many years. Maybe then they do not accept easy answers to these questions, and maybe they feel compelled to study them for a while by themselves, looking for their own answers.

Maybe they get used to studying these issues, and not feeling satisfied with superficial answers. Maybe they extrapolate this attitude to their relation with their profession, with society, and with themselves.

They may simply conclude that their teacher is a little stupid, though. One has to take some risks in life.

I was a relatively good student in high school, so I had some positive feedback from my teachers, and that made me feel well about my intellectual capabilities. Thanks to that, I managed to enter a good university for my undergraduate course. At the university, I was surrounded by colleagues who had been just like me in high school. As a consequence, I turned overnight from a reputedly intellectually gifted bloke to yet another average student in the crowd. Mind you, a crowd from where I could spot *really* intellectually gifted blokes, who were invariantly incredibly brighter than me!

Something seemed to be going wrong. I started suspecting it after my first lectures, and became certain about it after my first exams. My first reaction was to think that there was something wrong with the university, and that was the first time I reflected about the role of the university, science and technology, as well as why our society and economy were organized the way they were. It was escapism, I know, and perhaps this was a good indication that these reflections would not go too far. Luckily they did not, otherwise I could be turned into a fake philosopher and live my life running away from my weaknesses and frustrations.

Then more exams came, and I decided I would have to be more practical to survive in that university. So I relinquished to the common understanding for those issues and carried on. Without much reflecting, I managed to get my degree, and then to get a job, and then to do an MSc part time – when I started studying Artificial Intelligence – and then one day the questions came again, like ghosts from a forgotten past.

What is the role of the university?

What is science?

What is technology?

How do these three relate to each other?

Why are science and technology such strong drivers in our present society and economy?

That was when I decided to turn my career to academia, and the first step should be to apply for a PhD program. I was working with computer science by that time, and I enjoyed learning and playing with mathematics. I started reading what I suspected that could help me find the answers to my haunting questions. Many years later I realized that my unsystematic readings were mostly related to philosophy of science and philosophy of technology.

I had the intuition that I could get closer to my answers if I went to study in a good university, and again I had to be practical. In order to be accepted for a good PhD program, I had to make good use of my background, so I had to find some way to put together my professional experience, my previous studies of Artificial Intelligence, my interest in mathematics and my genuine motivation to do a PhD, which were to find answers for my long lasting questions.

I ended up doing a PhD in Artificial Intelligence in one of the top universities for this field, exploring logical and mathematical issues related to the modeling of human reasoning. Those were pleasant days indeed. Again I was surrounded by people much smarter than me, but I had got used to that and this time I managed to enjoy it. I was lucky enough to meet some of my best friends during my PhD years, and alas! Most of them were troubled by the same questions I was.

I slowly learned that it was not really the answers that mattered, but the questions themselves. There are no fixed answers to these questions; it is the process of permanently building and rebuilding answers to them that makes them useful and important.

And, by the way, those were not the only important questions whose answers should be reconstructed every morning.

I finished my PhD and got a position at the university in which I am still working, as a lecturer and researcher. How fine! I could finally focus completely on the issues that had pleasantly turned from haunting ghosts to beloved companions!

And so I did.

For a few months, perhaps.

The academic department that hired me has some lifelong internal disputes between research groups that strive for power, visibility and (what else if not) funding. At first I truly ignored this, then I tried to stay away from it, and then somehow I got mysteriously involved in it, also striving from power, visibility and money like a caveman fighting for food and water.

It took me many years to realize that some change had occurred. I was publishing fine and teaching fine but the spark was no longer there. And indeed I would probably not get out of this day-in-day-out teach / write / stab / get stabbed / home / sleep / teach... without the help of beloved friends.

It was my wife who one evening questioned me. She reminded me of when I arrived late from work, enthusiastic and cheerful, talking nonstop about things quite distant from her own experience and background in fine arts, and obviously happy. She compared that me with the other me she had in front of her, frowning and moaning all the time, sometimes talking about plots that were being planned against me, other times about plots planned by myself. She asked what had happened, and I was shocked to notice that something had happened after all.

Something that did not have to happen.

When we are children we are innocent and naive. Then we grow up, and the lucky ones loose both innocence and naïveté (the unlucky ones lose their innocence, and stay naive). Could we not only stop being naive, and keep our innocence? Are we really seeing *more* things as we grow old, or do we give up perceptions and purposefully blind ourselves to certain possibilities in order to get by?

Am I doing something useful with my work? Am I going round in circles? Do my papers do any good to anyone? Why writing them? Who reads them? And why do they read them?

Am I doing any good with my lectures? With my questionings? Does it make any difference to anyone, including me? If it does, what difference can it make?

Can I be truthful to my quests and to myself, and continue to work in this department? In this university? In any university?

What are universities good for?

What is science?

What is technology?

Why our society and economy are organized this way?

What makes all this *real*? What is real, and what is unreal?

I feel real good this morning. Some stuff I had learned in the past I have finally come to the conclusion that were dead wrong, and the feeling it brings is so good that I believe to have gotten it right this time.

I was told once that one *had* to get involved into office politics to survive in any organization, and therefore to accept the rules of the organizations, to take part in plots, to be deceitful and hypocritical, at least to some degree. As a corollary, in order to be successful in an organization one would have to give up their own views of the world – at least regarding to the business of the organization – and accept the organization's views.

Dead wrong. One *can* do all that, and there is nothing wrong about any of the above, but most certainly one does not *have* to do any of that. It is a matter of inversion. An organization is made of people;

therefore it is people that give reality to the organization, and not the contrary. Once you enter an organization, the organization is fundamentally changed. *You* create the organization, including its views and rules of behavior. Sure, when you start working as a trainee in the commercial office of Coca Cola in northwestern Brazil, it may take some careful analysis to sense that you are changing the whole of Coca Cola, but you are.

You can be disruptive or constructive, and the effects of your participation in an organization can be noted in the short term or in the long term. If you simply do not fit, then you are fired and your participation is short term disruptive: you did become part of the organization after all, taking responsibilities and all that, and you did promote changes in the organization by interacting with other people, so when you leave it is a part of the organizational body that is being taken away – an arm, a leg, half a hair or a tumor, in any case a part not to be neglected.

If you do fit and accept those rules I once thought were necessary, the effects of your participation in the organization have the potential to be – together with that of your colleagues that share the same attitude – long term disruptive: you are contributing to the creation of a self-devouring senseless monster, that one day shall implode under the weight of its own complexity.

To be constructive, you must feel responsible for the organization. You are, anyway. Are you used to complain about the government? You are responsible for that, too. About your family? About your salary? Global warming? Bad distribution of wealth? They are your responsibility. You may accept this simple fact of life, or you may not. It does not matter. You are still responsible for all that either way.

Let us say you assume your responsibility. Let us say you decide to do this when you take your job as trainee at that Coca Cola's branch in Brazil. You are the trainee, and yet you consider yourself as responsible for what Coca Cola does as the firm's president. You will certainly want to know how the organization works, and why it works that way. You will naturally wish to be constructive, accepting what there is that is good at the organization and trying to find ways to change what you consider that there is that is not so good – probably by discussing it with your colleagues, with your director, and so on. It is not likely that the changes you suggest are all going to happen. Actually, it is quite likely that many changes you suggest are not so good after all. But the willing to change, and the awareness of the possibility of promoting change, is where the real power lies.

Your interest and good will may resonate in the local organization with which you directly interact, and in this case it is likely that your participation in the overall organization is long term constructive. It can also happen that it *does not* sound at all like music in the ears of your colleagues or superiors, and in this case you may not fit and be expelled by the organization. In any case, you will have presented a possibility to the organization itself. Your participation will be still constructive, even if short term.

The relationship between individuals and organizations is obviously a two-way relationship. Just like the participation of an individual in an organization can be disruptive, constructive, short term or long term, the influence of an organization in the constitution of an individual can have all those effects.

Universities are organizations. They are organizations of a special sort, since they are *designed* to influence people in the long term. Undertaking a position in the university and not acting constructively seems to me, therefore, one of the most vicious possibilities of participation in our society.

However, how to determine whether some action is constructive? Who is to define the parameters to evaluate?

For an action to be constructive, something must be necessarily *under construction*.

Do we construct the reality in which we live? If so, do we live by illusions? Artificial worlds, to which we commit and which we accept to treat as real?

For what reason – or reasons – would we do that?

It is nearly time to leave to the university now. I shall be teaching programming to my students. Essentially, computer programs – of any sort – are machines that transform data. They are useful when they perform transformations that make sense to their users. What provides meaning to data and their transformations is some mapping from those data (and transformations) to a *model* of the world (or, more precisely, a model of some isolated phenomena that can be observed in the world).

In order to make sure that a computer program is correct, we must check if the transformations it performs correspond to what is expected in the corresponding model. Therefore, the correctness of a computer program can only be verified if its corresponding model (and the dynamics of the model) can be analyzed. Hence, computer programs are useful when they refer to models whose components and behavior can be checked, and thus observed. Such models to a large extent coincide with the commonplace notion of *scientific models*; hence it should be no surprise that computer programs have become such important and ubiquitous tools for scientists in the most diverse areas of scientific inquiry.

Now, if computer programs are machines to manipulate data, and there is a correspondence between computer programs and what scientists do with scientific models, could it be that scientific development is essentially data manipulation?

How are models built? Would it be the case that, in order to understand something, we would necessarily first have to blind and detach ourselves from other things?

What is science?

What is technology?

What is the role of the university?

How do these three relate to each other?

Why are science and technology such strong drivers in our present society and economy?

Tough questions indeed. Hard companions to keep, these ones.

Anyway, it is now time to go meet my students.

Again.

Too much thinking and no administrative work done.

Again.

Off we go.

Again.

Chapter 1

Introduction

"I", came a metallic voice from inside the closed helmet, with a slight echo as if it were not a throat but the very armor itself vibrating, "am Agilulf Emo Bertrandin of the Guildivern and of the Others of Corbentraz and Sura, Knight of Selimpia Citeriore and Fez!"

Italo Calvino

Our goal in this book is to present our conceptualization of information and of knowledge, and to clarify the distinctions between these two concepts. Indeed, they are more than concepts. These two words convey a plurality of meanings and have a deep character that goes beyond pure conceptualizations. By emphasizing their distinctions, we do not try to set them apart. On the contrary, we connect and relate them, whilst avoiding reductionist projections of one to the other and potential confusions resulting from the overlapping of their meanings.

We believe this can be useful for computer systems designers and programmers, especially those designing and implementing complex information systems and computer products in general. The distinctions between information and knowledge can help a designer to increase his/her awareness of the process of creative abstraction inherent to any design process, thus helping to situate and integrate the design activity in a wider context and a wider perspective of the concrete world that hosts the products of design.

Our primary goal is to present our personal views and reflections about these subjects to the philosophically minded inquirer. Our interest is to promote, through a better and deeper understanding of the notions of information and knowledge, a better and deeper critical understanding of information technology as situated in the full range of human activities, assuming as a principle that this range of activities cannot be properly appreciated when it is reduced to the simplified means-end schema proposed by Technology. We invite the reader to build his/her own points of view about these notions, considering our propositions as a starting point for a critical analysis and discussion of these points. With that, we believe we are contributing to a better understanding of the impact of technology – and particularly of Information Technology – in everyday life. Ultimately, a reflection on knowledge and information should help in the construction of a better understanding about oneself, which is a necessary step to a more balanced evaluation of Technology, its real contribution to our needs, its promises – both fulfilled and unfulfilled – and its relative position in our lives. This is the path we envisage to emancipate us from the grips of Technology, avoiding absolute dependence as well as an equally negative rejection of it.

We relate very closely the notions of *information* and *organization*. These two notions are not the same, but there is a bivalent relation between them – the more organized a system is, the more information it carries.

Information, according to our view, is also related to *complexity*. More complex systems can carry more information than the simpler ones. Or, to put it differently, a more detailed perception of a system, which reveals more of its complexities, carries more information about it than a less detailed perception.

A system made of a single component cannot be disorganized. However, it is so simple that it cannot carry much information. A system made of two components, which are however completely inarticulate with respect to each other, cannot carry much information either. A third system made of two components whose combined behavior is meaningful, i.e. such that the state of the first component is somehow *enriched* by the state of the second component, in such way that one component adds to the other thus contributing to the set of potentially meaningful states of the system as a whole, is capable to carry more information than the first two systems.

The more components are perceived in a system, the more information it can convey. The more articulate these components are, and the more important each component is to the behavior of the system as a whole, the more information it can convey too.

Through information, a person can filter selected aspects of the reality by choosing to perceive different views of this reality, through which more or less components are revealed, as well as their articulation and contributions to the way reality presents itself. Through analysis and then synthesis, we conceptualize and build abstract information models, whose objects of description are in this way put to our service. Information is useful to manipulate things as well as to design objects and artifacts to fulfill our needs. There is a risk, however, in the information based approach to reality. Namely, one can forget the abstract character of information and reduce the whole meaning of the reality to what is seen through information structures, thus forgetting that the constitutive meaning of reality goes beyond its utility. The constitutive meaning of reality is revealed and can be appreciated through knowledge, through which is unveiled the irreducibility of the world to rationalistic representations and reasoning.

Through information, a person can also build artificial and personal (information) structures based on these filtered aspects, and then perform "practical reasoning" – e.g. for problem solving – with and about these structures. Information is, therefore, an important supporting pillar for our technological society. It is difficult to decide whether information has been adopted as the foundation for our social, cultural, technological and economic organizations, in which case information structures and personal information structures have been the pre-requisites for the construction of our society, or our social, cultural, technological and economic organizations have driven our view of information as the means to approach reality, thus suggesting the development of personal information structures as conceptual devices for communication, in which case the perception of information structures and the construction of personal information structures are a consequence of our society – a "chicken and egg" problem.

Information – and the approach to reality through information and personal information structures – is definitely the foundation for classical scientific reasoning. In scientific reasoning, scientists approach reality with the concrete goal of extracting from it observable regularities that can be described linguistically in some unambiguous way. To do so, they select a portion of the reality to be observed, i.e. they select a certain *granularity of perception* through which they are able to appreciate a specific information structure from the multitude of information that is offered to them by reality.

When a scientist is a good observer, and/or when she is lucky, she manages to choose well, and selects an information structure that can be well represented using the available conceptual tools, linguistic tools (natural language, diagrams, mathematics, and so on) and technological devices (written language, computers, the Internet, etc.). She then builds a second information structure – her personal information structure – to represent the observed information structure found "out there" in the reality.

Unavoidably, the personal information structure will be contaminated by the scientist's presuppositions, personal goals and ambitions, prejudices, and so on. A skilled scientist is capable of minimizing the interference of this contamination in the construction of the personal information structure, thus building artificial information structures that are faithful to the originally perceived ones.

The personal information structure is then published as a scientific model, and proposed for adoption as description and explanation of the reality that provoked its construction.

Good science results from – among other things – perceptiveness, skill to build as neutral as possible descriptions and explanations, and of course a pinch of luck. We shall avoid the temptation to discuss in this book the purposeful and objective behavior that we believe that brings luck to a person, as we do not feel prepared for the moment to discuss with the necessary clarity the presuppositions we believe that must be unveiled, analyzed and perhaps discarded for that. For the moment we shall focus on acquired skill and perceptiveness.

It does not surprise us, therefore, after portraying the activities related to scientific discovery the way we did, that technological devices to store, process and disseminate information structures have become such prominent tools for scientific discovery, modeling and development. Among these devices, the computer (in all its flavors and sizes) is a very remarkable technological achievement. Prior to that, classical mathematics is the most remarkable conceptual and linguistic tool produced by our society.

Indeed, we should be surprised to consider how long it took for pure mathematics and computational techniques to have spread outward the boundaries of Engineering to reach diverse scientific fields such as Biology, Linguistics, Social Sciences and Economics.

As we will develop further in the chapters to follow, the prominence of computers and computation in modern scientific thinking and in all sorts of activities in a society shaped and ruled by values proposed by science and technology, is a very natural and expected situation. Computers, after all, are essentially and fundamentally devices with the following capabilities:

- Efficient storage and processing of information structures.
- Effective presentation of information structures, in such ways that can impress our senses remarkably vividly, thus bringing "concreteness" to abstract structures that become then part of the perceived reality, instead of just invisible shadows of that reality.
- Effective communication of information structures, providing means for novel forms of interaction among people.

It is hard to conceive a device that responds more perfectly than computers to the requirements of a society that craves for interacting through personal information structures.

Information gives room to very simple notions of knowledge. As we develop in the chapters to follow, our proposed notion of *knowledge* relates to *identification* – a person can say he knows a beloved friend when he somehow is capable of "standing on his friend's shoes" and standing on his behalf in any given situation; a person can say she knows a place when she somehow is capable of integrating with that place so tightly that any change in the environment also changes the person, and vice-versa. Clearly, through information structures a person can only know information structures – most commonly personal information structures, more rarely the information structures found in the reality. Knowledge of the reality – natural reality related to a most beloved place (for example, the backyard where one used to play as a child), to others (for example your son) or to oneself – necessarily bypasses information structures (Figure 1.1).

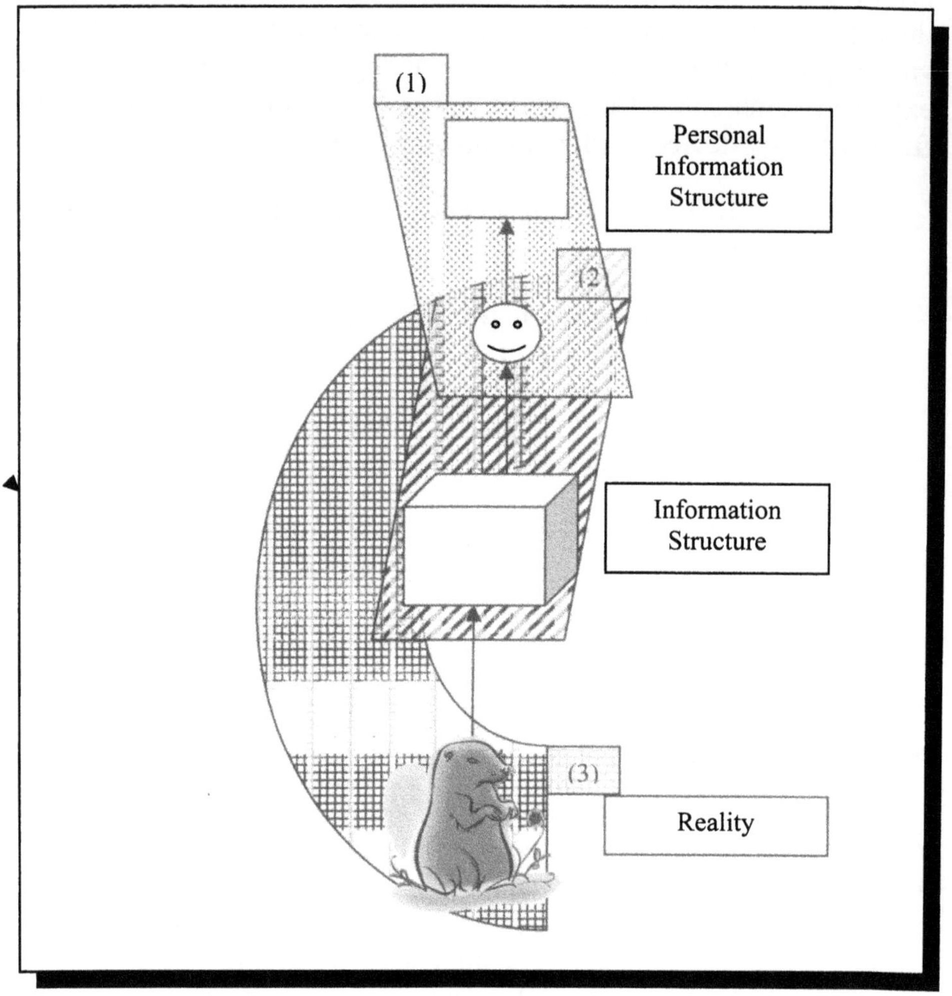

Figure 1.1. Knowledge and Information: a person can (1) *know* a personal information structure, (2) *know* an information structure observed from the reality or (3) *know* reality itself, when the relation with reality is not mediated by information structures.

In the final chapters of this book we discuss a little further – although without resorting to formal or mathematical structures, and hence proposing no operational model for our proposed concepts – our views about knowledge. We also suggest – and this may be the main reason for us to write this book – that this general notion of knowledge, together with the universal quest of Man for knowledge, constitute the driving force that precedes, creates and explains our social organization and our information based relation with reality, in which information structures are a tentative shortcut and a mediator in the relation with reality.

Intended audience

Our goal in this book is to reach the widest possible audience. For that, we have carefully avoided mathematical formalization. We have also limited quotations and bibliographical references to a minimal. Finally, we have avoided technical language. To be sincere, we have written this book the way that was most pleasant to us to write. We have also considered during the preparation of the book the possible way to write it that would make it most pleasant and amenable to read by the largest possible amount of readers.

After experimenting with a large number of forms to present our ideas, we have decided for a "conversational" style, in which we as much as possible simulate a dialogue between you – the reader – and us – the authors. The contents of this book have been built through dialogues between us – Jaume and Flávio – every time an opportunity came (mind you, Flávio lives in Sao Paulo, Brazil, and Jaume lives in San Celoni, Spain). We managed to meet approximately once a year, for two or three weeks in a row each time we met, when we devoted ourselves to *conversation*. We are proud to say that we have managed to be very productive and focused in our conversations, but that was all we did when we met: we talked about the subjects of our interest, which we now share with you along these pages.

We tried as much as possible to re-create in the text the atmosphere of leisure and excitement we found in our dialogues. This is the reason we decided for the slightly unusual form of presentation of our ideas you find in this book.

We understand nevertheless that texts following more closely the tradition of scientific writing must come together with a text like the present one, in order to facilitate the extension of this work by other authors, to give room to the application of some of the ideas presented here for practical problem solving, and to reach some specific, technically minded audiences. We have organized these texts as technical articles, which have been published in specialized technical conferences and journals, and in which you find the usual and expected mathematical rigor and appropriate specific links to existing bibliography. We have also ventured ourselves in the design of applications of some of our ideas, which have been published also in technical conferences and journals, as well as in the form of ready-to-use computer software. The software applications we have developed based on the ideas presented in this book have been tools to augment human capabilities to communicate, to solve problems, to design artifacts, and for people to interact in a very wide sense with each other and with the environment.

The software applications we have built – and which we would expect other people to start building influenced by our ideas presented in this book – have been built for purposes previously identified and attained by existing software products. We expect that the perspective proposed by the concepts and ideas discussed here influence the design of these systems, thus enabling the construction of more interesting and useful products. The ideas presented here have permeated and influenced our production as a whole, and therefore every software system, conceptual construct or scientific model we have built recently has been conceived through the perspective proposed in this book.

More specifically and directly, we have designed and programmed systems in which state-of-the-art software design techniques, such as those found in multi-agent intelligent systems, distributed problem solving and augmented reality, are put at the service of enriched communication among people for collaborative problem solving. We have developed computer games, knowledge management systems and learning management systems to support distance learning as well as presence-based learning using these ideas.

Chapters' contents

In chapter 2 we present our conceptualization of *information*, as the measure of organization of an observed portion of reality. According to our view there is a notion of information, therefore, that is inherent to reality and independent of the observer.

In chapter 3 we introduce the notion of *information flow*. Information flows between intelligent agents, and therefore information flow depends on these agents to occur. What we refer to as information flow is the purposeful action of an agent sending pieces of information to a second agent, and that second agent receiving it. Information flow brings with it the notion of roles for intelligent agents, namely the roles of *sender* and *receiving* agents.

Information flow can be used for many purposes. A distinguished one is the grounding of *shared understanding*. In chapter 4 we discuss this notion. Essentially, shared understanding occurs when, through information flow, agents engage in a dialogue through which they asymptotically and collaboratively build a public information structure.

Shared understanding allows that agents coordinate their actions. This can be done so that each individual agent has an increased amount of available resources to reach their individual goals. It can also be done so that a group of agents acts collaboratively to solve a single problem. Collaborative problem solving is the subject of chapter 5.

Collaborative problem solving is an important and interesting issue, specifically for information systems architects and, more broadly, for any individual engaged in social interaction. Many engineering techniques have been devised in recent years to improve the effectiveness of designed information systems so that they promote collaborative problem solving. Among these techniques we highlight "ontological reasoning". In chapter 6 we briefly revise the basic notions of ontological reasoning and present our personal views about this information systems engineering methodology.

Social interaction, no matter how precisely circumscribed by social rules and protocols, always gives room to misunderstanding and ambiguity. Essentially, information systems are tools to promote social interaction, more often than not for specific purposes and with specific goals in mind. Uncertainty as a measure of ambiguities and of potential misunderstandings in social interaction is therefore an important factor to be taken into account in information systems design, implementation and utilization. In chapter 7 we discuss this point further.

Ultimately, social interaction should promote deeper and better mutual understanding among people. *Knowledge sharing* is the name we employ in this book for this notion of deeper and better mutual understanding that ideally should result from social interaction – and that therefore should ideally be the goal of every information systems designer. In chapter 8 we discuss the concept of knowledge sharing.

Finally, in chapter 9 we discuss our proposed concept of *knowledge*. To put it briefly, knowledge is a "something" that occurs when the knower feels integrated with the known. Knowledge sharing is, according to our view, a path to knowledge.

Chapter 10 is left for conclusions and some final discussions to wrap up our propositions.

Intermission

Between each chapter we have added an intermission.

In the showing of a film, a theatrical play or an opera, an intermission is a break between two parts of a single piece. They are included for technical reasons (e.g. so that the actors and singers can rest, or the stage is reorganized for the next scenes), but they are also appreciated by the audience as a moment to relax, have a cup of coffee and socialize.

Our intermissions are the moments in which we intend to socialize with you, dear reader. Essentially, in each intermission we focus on the subject of the previous chapter and reintroduce it in a more playful manner. As stated previously, we intend as much as possible this book to be perceived as a dialogue between you – dear reader – and us the authors. If we were indeed talking face to face, or if you were in a conference room interacting with us as if we were the speakers in a seminar, we would certainly have a coffee break in which we could discuss in more relaxed manners the subject matter of our dialogue, relate it to other subjects, make some jokes, and so on.

Each intermission is itself the chapter of a short story. This is a fictional story, although slightly related to recent events we – Jaume and Flávio – have lived in our personal lives. It is completely up to you, dear reader, to choose the way you prefer to read this book: you can read all the chapters first – in sequence, of course – and skip all intermissions, and then read the intermissions afterwards, or you can read all intermissions first and have some fun, and then read all the chapters, or you can read the whole book page by page, thus reading a chapter, then the corresponding intermission, then the next chapter, then the next corresponding intermission, and so on.

These intermissions are intended for you, dear reader, so that – as mentioned before – you and we can socialize, chat a little and sip a cup of coffee between technical discussions. They also worked as intermissions for us (the authors) while we were preparing the book, quite like those restaurants we find along highways and where we so pleasantly can stop to stretch our legs for a while before we continue on a long driving trip.

As you are certainly well aware of, many important things occur during theatrical intermissions – marriages break, romances begin, business transactions are effected, and so on. Many important things occur during our intermissions as well. Indeed, these are the points in the book where we more freely expose our ideas and ourselves.

Some day we may be brave enough to write a book containing only intermissions. We will find the way to let you know when this happens.

About the quotations in the beginning of each chapter

The quotations in the beginning of each chapter all come from Italo Calvino's *Il Cavaliere Inesistente* (1959 – English translation *The Nonexistent Knight,* Hartcourt Brace, 1977).

This fine small jewel of modern literature conveys in artistic form in between its lines many of the concepts we have developed in this book. We strongly recommend the reading of this novel to complement the reading of our book.

Chapter 2

Information

World conditions were still confused in the era when this took place. It was not rare then to find names and thoughts and forms and institutions that corresponded to nothing in existence. But at the same time the world was polluted with objects and capacities and persons who lacked any name or distinguishing mark. It was a period when the will and determination to exist, to leave a trace, to rub up against all that existed, was not wholly used since there were many who did nothing about it – from poverty or ignorance or simply form finding things bearable as they were – and so a certain amount was lost into the void.

Italo Calvino

We start with a conceptualization of *information*. Notice that we avoid stating that we are going to provide a *definition* of information. Many accomplished scholars have endeavored on this errand, and the study of their work certainly leads to the conclusion that, to say the least, this is a complex task.

A definition leaves no doubt about the nature of what is being defined. It sets apart the defined object from every other possible object, so that when that object is evoked – say, through its *name* – every possible feature of that object comes with it, and there is no risk at all that a feature that does not refer specifically and solely to that object also comes with the object.

A conceptualization is a much weaker and permissive notion. When we say that we have built a concept that describes an object, all we are saying is that we have managed to capture *some* features of that object, as well as we may have managed to identify some *non* features of the object. When we evoke a concept, a collection of objects steps forward, from which any individual object can be taken. The distinguishing features that join together all objects in the collection comprise the set of shared features identified in the concept, which do not preclude the objects from being different from each other, due to features not considered in the conceptualization.

In a more precise and synthetic way, conceptualizations are *polysemic*, whereas definitions are not. We can produce a definition for the mathematical concept of natural number, for example, but we would hardly be capable of producing a precise definition for the concept of Life, considering how this concept has been employed in everyday discourse. It is definitely not the same as saying that a definition of Life does not exist, though. Whatever this definition is, our problem is that we have not reached as yet the point of synthesizing it in such way that everyone would be happy with the definition.

Previous conceptualizations of information can be classified in two broad groups:

1. The *subjectivist* approach to the conceptualization of information. According to this approach, and paraphrasing the well known popular saying, information is in the mind of the beholder. This view supports the "data / information / knowledge" hierarchy often employed in organization theory. According to this view, the world is made of data. Any event, natural or man-made, is essentially the means to produce data. *Data* are purely objective and do not depend on any external entity to exist. When data bump on an *intelligent* entity and make sense for that entity, then that entity builds *information*. Information, therefore, depends on the existence of external data as a stimulus, but is built entirely within those intelligent entities who are subjected to data. Information is interpreted data, and interpretation depends on the interpreter, who is in this case a *situated* intelligent agent. There cannot be information without an agent to interpret data. Finally, *knowledge* is actionable information, i.e. information that can be turned into action by the situated intelligent agent. Knowledge is the key to purposeful transformation of the world by an agent. It is therefore dependent on information as well as the goals and capabilities of the agent who has produced that information, based on the received data. It is not difficult to notice the utilitarian motivation underlying these characterizations of data, information and knowledge. These are indeed the characterizations promoted by businessmen to build rational models of interaction for effective business transactions.

   Data derive from the external world, to which we refer as *reality*. It is not equivalent to reality, since only selected facets of reality can be captured in the form of data. Data, therefore, capture just a part of reality. Information is built by an intelligent entity, who constructs an organization for data in order to make sense of it. The subjectivity of this approach comes from this point: the originally unstructured data is given structure by the intelligent entity, thus producing a new concept named information. Knowledge is information that prepares the intelligent entity for action. Action transforms the reality, hence closing the circle (Figure 2.1).

2. The *objectivist* approach to the conceptualization of information. According to this approach, information is "out there", regardless of having someone to appreciate it. The world is not chaotic, and the difference between chaos and not chaos is information. Information is the extent to which a system is structured and organized. For example, a molecule of a chemical compound *has* information, which is expressed in its organization and structure. It has information as much as it has constituent chemical elements. Graphite and diamond are both made of carbon, the difference between them being their internal organization and structure, hence the information they carry. A complex system has more information than a simple one, and the amount of information that constitutes a system (together with other elements of different sorts that are also constituents of the system) can be even quantified. Information according to this view is, however, *meaningless* and *purposeless*. It takes an agent to capture information, interpret it and make use of it to act and to reach goals (Figure 2.2).

Our conceptualization has a pinch of each of these groups. It is not an "average" position between the subjectivist and the objectivist approach, nor is it an encompassing view to cover these two approaches. We are definitely not interested in "average" views, which in this case would mean adopting an unclear and not well defined position – it would be, in a sense, like choosing to be half way between a salty and a sweet salad dressing, ending with a tasteless sauce. As for an encompassing view, we would indeed have liked to produce one, but this is beyond our competence.

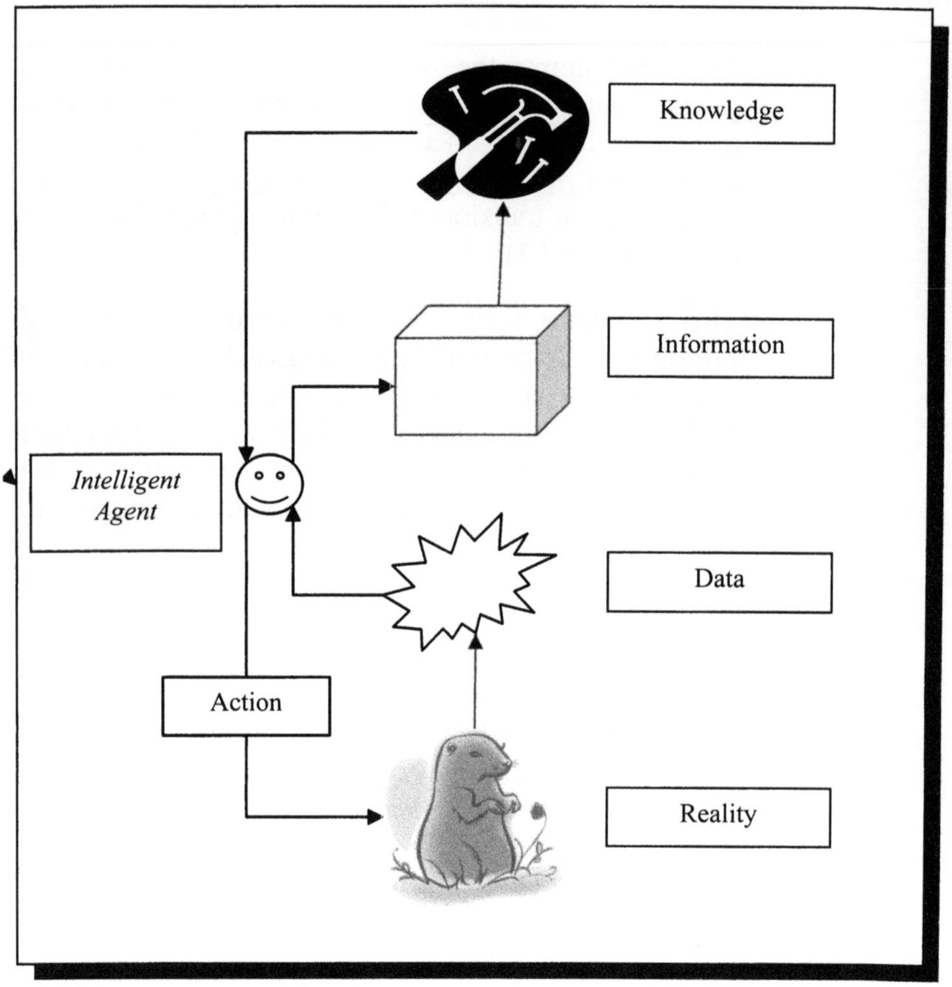

Figure 2.1. The subjectivist approach to information

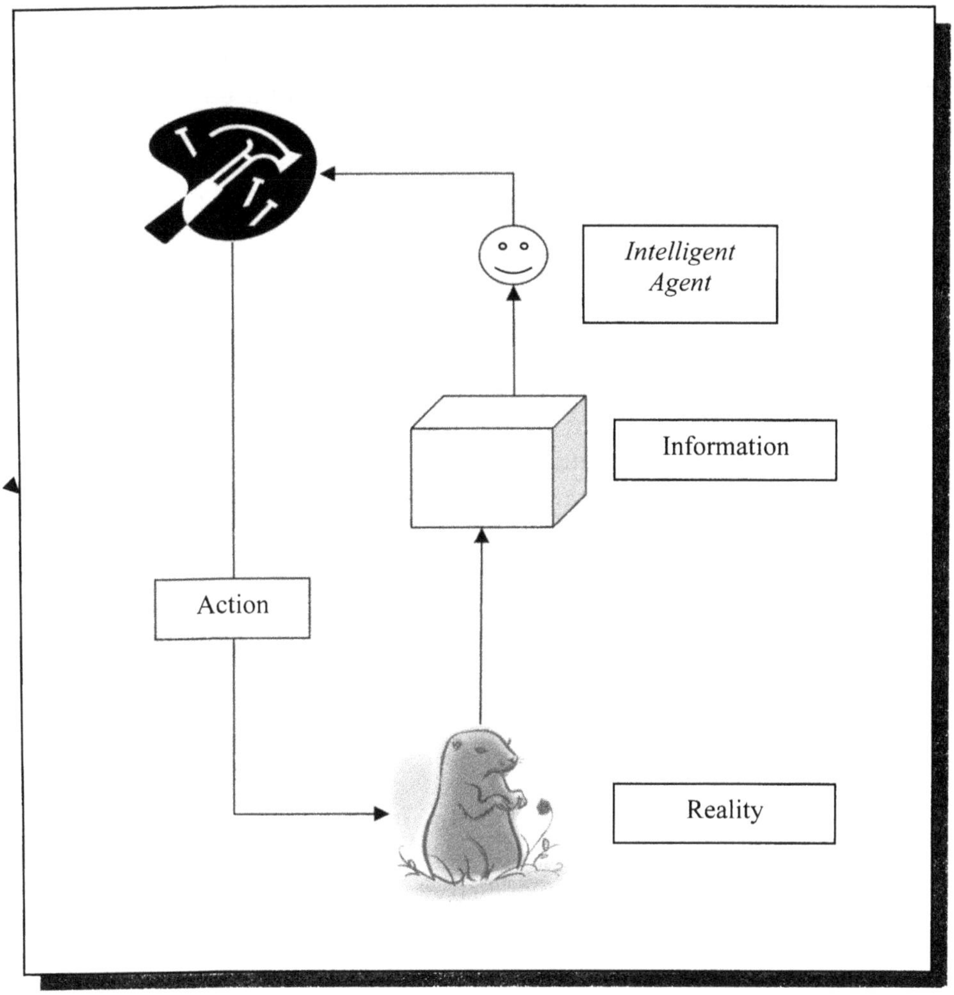

Figure 2.2. The objectivist approach to information

Our approach is that of a *chef*: we have treated these approaches as fine spices, which we have tried to put together in a balanced way so as to create a tasteful sauce that we hope will please those who taste it. As any sauce, we expect this one to be good for certain meals and inappropriate for others. Personally, we like the taste of the sauce we have prepared, and this is why we are sharing it with others through this (cook) book.

In other words, we have organized a conceptualization of information that has suited our purposes very well. We have used it as the foundation of the recent work we have developed, as well as a conceptual framework to understand the work of others, and this is the reason we have decided to share it with others. We do not claim it to be a general and all-encompassing characterization of information, as we do not believe it is. It has proven to be quite general, though. We will be happy to learn from the readers about challenging problems and situations for which our conceptualization proves to be inadequate, so that we continue to perfection it.

Agents and the world

In our conceptualization, information is the means through which situated intelligent agents most frequently relate to the world. Before we continue, it may be worth clarifying what we mean by *situated intelligent agents* and by *the world.*

The agents are our focus of interest. Essentially – and this will be clarified along the book – they are *people*. No robots, knowledge-based systems or synthetic virtual characters, but *humans*. People are by definition intelligent, although they can switch off their intelligent behavior when they find it appropriate – and this will also be clarified in the following chapters. With that we mean that we will not worry so much about precisely defining *intelligence* – whatever conceptualization that necessarily includes humans as intelligent agents will do for our purposes, and nearly all commonsense interpretations of intelligence follow this requirement.

We continue to call our agents "*agents*" instead of simply people, however, so that we can consider some variations on the theme later on. For example, agents can have *proxies*, which strictly speaking should not be called agents themselves. It has been common practice in Artificial Intelligence, however, to name "agents" these proxies, thus forming the basis upon which the theories and systems for multi-agent systems have been built. Strictly speaking, according to our view and to our proposed terminology, these are in fact multi-proxy systems.

Our agents are temporally and physically *situated*, hence it does not in principle suffice to identify a person to build expectations about the way that person relates to the world – we must identify a person situated in a specific temporal moment and geographical position.

In order to characterize the notion of the *world*, we adopt initially a rather egocentric stance, to be refined later. Each individual agent acknowledges the existence of two major entities in the Universe: the proper agent and whatever else that is not the agent, which is called "the world". This initial separation will however be challenged in the chapters to follow.

We assume that certain phenomena occur regardless of the will or the perception of any specific agent. These are the phenomena occurring in and belonging to the world. Chemical elements simply exist, thunderstorms simply happen, the price of oil in the New York Stock Exchange can be envisaged as simply oscillating, and so on. These phenomena simply occur, regardless of being perceived or not by an agent.

Following the objectivist approach to information, there is an inherent organization and structure to these phenomena. Where exactly, however, do we fix the boundary separating one phenomenon from the others?

The world must be actually comprised as a fully connected network of events. "Network" is actually not so appropriate a characterization to convey what we intend to mean, since it presupposes an understanding based on a very specific mathematical model, made of nodes and connections which is inherently discrete.

We provisionally borrow this concept, though, for the want of a better notion to express what we wish to express.

Reality and structure

Our starting point, therefore, is this notion of a surrounding world in which events are occurring. The world is a fully connected network of events, a statement to which someone attached to chaos theory would concur. What the world is not, however, is fully chaotic. There is order in what happens.

The order, organization and structure of the world can be appreciated in their entirety by some individuals in specific situations, and more rarely by certain individuals nearly most of time. We conjecture that this can be the case, for example, of certain mystic ascetics and tribal shamans, who by means yet not fully explored by science somehow manage to integrate with the natural world – including here facets of natural reality that may pass unnoticed to most of us – in such way that the boundary that separates them from the world is blurred.

The more common situation, however, is this order to be only partially perceived by individuals. To make an analogy, television channels emit electromagnetic waves all around. A TV set can be adjusted to a specific range of frequency, and then capture what is being broadcasted by one specific channel. What appears in the TV set, however, is not what is going on in the studio, but only selected scenes to be shown to the audience. The reality of what is occurring in the studio – the world – is only partially captured by the cameras – the perceivable world.

Our analogy breaks here, because in the case of the TV channel what is shown in the TV set is purposefully determined by the TV channel managers and technicians. In our case, there is the world purporting a collection of structured events to the individual's appreciation. The world is, in principle, purposeless, in the sense that what is presented to an individual is not *a priori* filtered depending on who this individual is. What the world presents to the individual we call *reality*.

We must observe here an underlying presupposition we accept in this work: reality is itself a name we are adopting for appearances proposed by a further and deeper notion, here called *the world.* We shall not discuss directly the world in this book. Our discussion shall start from the reality onwards heretofore.

In the case of the TV channel, even though the scenes shown through the TV set do not capture the entirety of what is going on in the studio – the *reality* of the studio – there is perfect correlation between events shown through the TV set and events occurring in the studio. Likewise, reality opens up for an individual a multitude of perceptions, which are correlated to events in the world and somehow capture facets of these events and how they are related to each other.

The individual does not construct these perceptions, they are inherent to the world and how it is organized. They are, nevertheless, separate systems of events. We call each of these systems an *information system* (Figure 2.3).

Information systems

An information system, therefore, in our conceptualization, is an organized network of abstract entities – which we simply call *pieces of information* – that is correlated to the reality. This correlation can be perfect, in which case to each noteworthy event in the information system corresponds one or more noteworthy events in the world. It should be noticed that it is not necessary that to each noteworthy event in the world corresponds an event in the information system. This is indeed the essence of our conceptualization: an information system captures a *partial* view of the world, even when the correlation is perfect. Hence, the information system leaves behind certain facets of the world, and puts together as indistinguishable events other facets of the world, namely those ones corresponding to events that are related to a single noteworthy event in the information system (Figure 2.4).

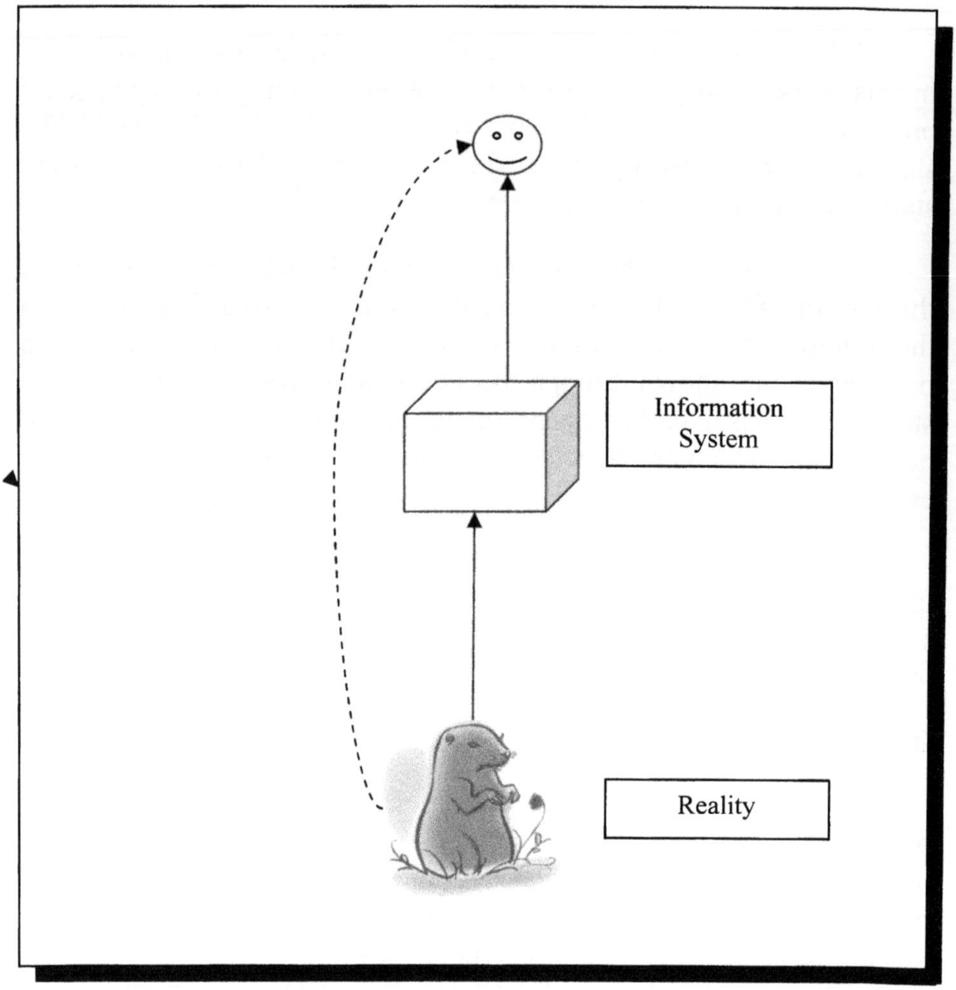

Figure 2.3. Information systems

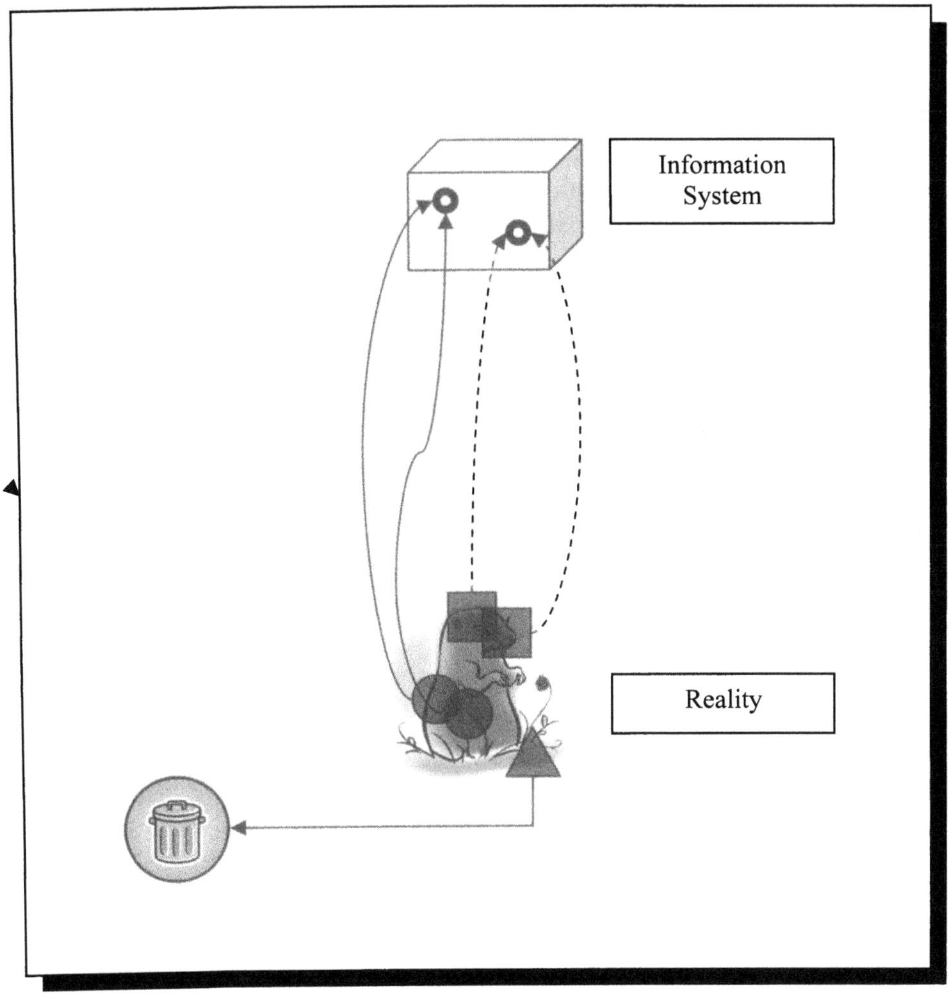

Figure 2.4. An information system. The squares and circles are put together as indistinguishable events and the triangle is dismissed as irrelevant (and sent to the trash bin)

As a concrete example of an information system in which the correlation between events in the information system and events in the reality is perfect, consider the field observations performed and recorded by a skilled botanist. The botanist can be interested in counting the number of specimens of a certain plant in a certain region, such that a specific malformation in the leaves can be found. For this botanist, all plants in the world will be classified in three disjoint groups:

1. The plants outside the region of interest.
2. The plants within the region of interest whose leaves do not present the observed malformation.
3. The plants within the region of interest whose leaves do present the observed malformation.

The assumption that the botanist is skillful is added here precisely to assume that the classification is perfect. Nevertheless, individual plants are grouped together in equivalence classes. The observation of an individual plant with the observed malformation relates to a noteworthy event in the information system containing only these three disjoint sets of plants, instead of individual plants. To the noteworthy event associated to the statement "observation of a plant with the malformation of interest in this study" corresponds any individual plant presenting that malformation. There is a collection of events in the reality, therefore, associated to a single event in the information system.

The correlation can also be imperfect, in which case to a noteworthy event in the information system there *can be* a corresponding event in the world, but a certain degree of uncertainty must be taken into account. As a concrete example of information system in which the correlation between events in the information system and events in the reality is imperfect, consider the observations performed and recorded by a skilled meteorologist. No matter how precise are the observations and how skilled is the meteorologist, to each observation recorded in the information system there can be a corresponding event in the reality, but, as we are all well aware of, meteorological predictions and observations are hardly devoid of a certain degree of uncertainty.

In the two examples above, we referred to an agent who would be recording the corresponding information system (a botanist in the case of a perfect information system and a meteorologist in the case of an imperfect one). We must stress our view that these agents are *recording* observations of information systems that were "already there", regardless of these agents.

Any selected slice of the reality offers a multitude of information systems to be perceived. Each information system captures and reflects some facet of that specific portion of the reality.

Each information system, therefore, has its own structure and organization, characterized in terms of its constituent entities and how they relate to each other. An information system relates to the world, more specifically to how the world is structured, which we call here *reality*.

Observation suggests that it is more comfortable for individuals to relate to information systems than to the world directly. For most individuals, information systems are the means through which they connect to the world. By definition, however, no information system conveys a complete rendition of the world. Information systems, therefore, present themselves to an individual with apparently paradoxical and conflicting roles: they at the same time provide the means for the individual to access reality and detach the same individual from the direct connection with reality, by acting as mediators that provide to the individual perceptions that correlate to facets of reality, instead of the fullness of what is going on in the actual world.

An information system can be linguistically represented in terms of an organized collection of structures of symbols. This collection of structures can be represented statically and exhaustively as a list or any other sort of diagram, in which all possible structures are explicitly represented *a priori,* or dynamically and on demand as an abstract machine comprised by subsets of structures and transition rules between these subsets. Intuitively, static representations relate better with static snapshots of facets of the reality, and dynamic representations relate better with dynamic facets of the reality which are better understood through simulation.

There are two natural notions of quality of an information system, related to how it correlates with facets of the reality and to the degree to which its events correlate to separate events in the world:

1. The *completeness* of the information system, i.e. how "wide" the facet of the reality it captures is, and

2. The *precision* of the information system, i.e. how close to perfection is the correlation between noteworthy events in the information system and noteworthy events in the corresponding facet of reality. A noteworthy event in an information system is either a specific set of structures of symbols or a specific transition between two sets of structures of systems.

Considering our previous example of the meteorologist, an information system *A* is more complete than an information system *B* if *A* includes more meteorological phenomena than *B*. The information system *A* is more precise than *B* if the weather forecasts based on *A* correspond to reality more often than those based on *B*. Notice that we could have, for example, two information systems *C* and *D* such that *C* is more complete than *D* but *D* is more precise than *C*. Moreover, the quality of two information systems can only be compared, evidently, if they refer to the same portion of reality.

Completeness and precision are inherent to an information system. The relative importance of each of these measures of quality depends on the agent that perceives the information system.

The paradoxical character of information systems becomes clear when we consider these quality measures. The more precise and complete an information system, the better its quality is. The perfect information system, therefore, would be a symbol system complete enough to capture all facets of reality, and precise enough so that to every event in the reality there would be one precise event in the information system. For an agent to relate to this information system it would be just as difficult as for the agent to relate to reality directly. An agent uses an information system because it is simpler to interact with than reality. Hence, the utility of an information system is entailed by its imperfection.

There is no need of a goal or a purpose for an agent to interact directly with reality. Typically, an agent may wish to interact with reality through information systems, however, for some specific purpose and in order to reach some specific goal. The alternative of interacting with an information system instead of with reality directly can be attractive for two reasons:

1. The agent may need *efficiency*: when an agent has a specific task at hand, for example, to be accomplished before a deadline and making use of a controlled amount of resources, it may suffice to work based on a simplified view of reality, which albeit incomplete and imprecise conveys the necessary means for the task to be accomplished accordingly.

2. The agent may need to *work collaboratively* with other agents: in this case, the agent may need to communicate to other agents his/her views of reality and the task at hand. Since information systems are symbolic representations of facets of reality, they can be transmitted between agents. This shall be studied in detail in the following chapters.

The presupposition for an agent to interact with reality through information systems is, therefore, that the agent does so in order to reach some goal, execute some task and satisfy some purpose. The agent must have needs to be satisfied to consider interesting this sort of mediated and restrictive relation with reality. This presupposition explains our conjecture about mystical ascetics and tribal shamans: these individuals are distinguished by not having egotist needs to be satisfied, and hence by being capable of releasing themselves from doing things, relating with the environment and with other people always for a purpose, bearing some goal in mind and in order to efficiently perform tasks. It is this distinguishing characteristic that enables them to unmediated relation with reality.

A non-rationalistic stance

As a first introduction to our view and propositions, we admitted some implicit presuppositions which we now unveil and generalize. Essentially, our presentation so far is aligned with what has been called the *rationalistic approach* to information processing.

Essentially, the rationalistic approach assumes that an agent and the world are independent of each other: events in the world occur independently of any specific agent, and the agent has a living of its own, independently of the world. The agent may *choose* to interact with the world, for example through information systems and planned actions with predictable effects.

This is not the stance we intend to adopt in this work. Our view is that all agents are fundamentally part of the world, *and vice-versa*, i.e. the world is fundamentally part of every agent.

Putting this in the simplest possible terms, we have stated previously that the world can be comprised as a fully connected network of events. We now add that the agents that perceive this world are also part of this network. Reminding that our agents are also situated, we have a highly interactive system whose components are correlated in such entangled way that they must all be considered at once. Move one agent from one place to another, or take it to be at a certain place in a different moment, or else replace an agent by a different one and everything in the system changes.

Abstractions of the reality in the form of information systems are strong simplifications in which it is assumed that certain sub-networks in the connected network devised above can be ignored. They can be useful for certain practical purposes, but the price to be paid by the employment of an information system instead of direct contact with the reality is high: when an agent accepts the interaction with an information system as a replacement for its participation in the world, it has to accept two consequences of this hypothesis:

1. It will no longer have direct access to reality. Instead, it will only interact with the abstraction of reality it selected, represented by an information system.

2. Considering that the agent has a perception of itself as a component of reality, and therefore not of an abstraction of reality, by interacting with an information system instead of with reality it will detach itself from the representation of the reality it chose to interact with.

The agent must be prepared to give up its perception on behalf of an abstraction of itself, e.g. as a pre-established set of roles and behaviors determined in accordance with the information system it has decided to use.

Our non-rationalistic stance precludes, except for highly controlled and oversimplified connections with reality, the possibility of detaching information processing from the information processors, thus disembodying the agents who perceive and process information, deeming them replaceable. In other words, our conceptualization of how agents relate to the world precludes a technological possibility so frequently explored in science fiction novels, namely that a human being will be able some day to transfer her essence from her body to, say, a robotic body or a computational database.

Bibliographical notes

The subjectivist approach to information and knowledge – and more specifically the "data / information / knowledge" hierarchy – has been employed in a large range of literature related to business intelligence and "knowledge" management. The reader interested in this perspective on the conceptualization of information and knowledge can find interesting material in the book by Tiwana (1999) and the bibliography cited there.

The objectivist approach to information we have employed in our work is influenced by Dretske (1981) – although our conceptualization is not fully aligned with Dretske's, especially in what relates to knowledge and how knowledge and information are linked to each other. It is interesting to notice how Dretske's work is becoming more influential nowadays to the analysis and design of systems for information processing and artificial intelligence.

The notion of unexpected indirect connections among seemingly unrelated events is at the heart of chaos theory. Chaos theory has evolved to a rather sophisticated branch of applied mathematics, which provides mathematical models to help understanding complex natural phenomena. We have borrowed from this theory the notion of unsuspected connections, to found our arguments towards the inseparability of agents and their environment. A recommended reading for those interested in a light presentation of this mathematical theory can be found in Stewart (2002).

The so called rationalistic approach to information and knowledge has been very much influenced by the Physical Symbol System Hypothesis (Newell and Simon, 1976), which stands for the disembodiment of reasoning. According to this hypothesis, reasoning can be fully understood as the manipulation of abstract, mathematical objects. Reasoning can be *implemented* in a variety of ways, such as biological agents like you and us or computer software. In this work we challenge this notion of disembodied information processing – and reject the notion of disembodied knowledge, as will be clarified in the chapters to follow. Other authors before us have followed the same path, of which we highlight the book by Winograd and Flores (1986), which has been very influential to ourselves as well as to many others. To them we owe the separation between rationalistic and non-rationalistic approaches to understand information and knowledge. Based on the distinction proposed by Winograd and Flores (1986), we have positioned ourselves on the non-rationalistic side of the street.

The rationalistic view, aligned with the Physical Symbol System Hypothesis, admits the notion of interchangeability of implementation of agents, thus admitting the possibility of human beings having their essence "downloaded" to a robotic embodiment or a computational database. This possibility is considered non-fictionally, for example, in the book by Moravec (1988). Moravec's work is, in turn, analyzed in the book by Hayles (1999), who argues – similarly to us – that the possibility of moving the essence of an agent from one embodiment to another can only make sense for highly controlled and simplified agents, environments and interaction protocols.

Intermission

Gigino fell down that tree when he was eight. He did not break anything, but got quite a few bruises. And as in many other occasions, nearly gave Dr. Agnelotti a heart attack.

Cicilla used to meet her boyfriend close to that newspapers stand, after school. Mrs. Agnelotti met them there once, and that was when Mrs. Agnelotti got close to a heart attack.

Time flows differently in San Isidro.

It is two weeks since I arrived here. There are times when it seems I have just arrived, and there are times when it seems I have lived here for many years.

Dr. Agnelotti kindly invited me to stay at his home during my visit to the Institute of Advanced Research. "This way we can chat on the way to the Institute and back. In the evenings we can go to the ice-cream shop. And of course you will save some money staying with us."

The spell of San Isidro and the stories of the Agnelotti family were not mentioned by then.

San Isidro is a very small village. San Isidro is also very ancient. Some remains of a roman aqueduct can be found within less then ten minutes walk from the central square. The local choir makes a presentation of Christmas carols in mid December every year under the ruins of the aqueduct. The concert, I have been told, is always very beautiful. I hope I can be around at the appropriate time some day.

Yet, San Isidro does not appear in any tourist guide. There it stays, quiet on the surface, unknown and unheard of.

San Isidro the forgotten.

San Isidro the proud.

The Agnelotti have their roots connected to the roots of the town. Dr. Agnelotti once summarized: "We move to study, we move to work. One day we move back. This is the way it has always been. For the obvious reason: can you imagine a better place to live?"

Dr. Agnelotti and his wife live in a comfortable old house very close to the central square. The house itself is not so big, but the back garden is lovely. And quite large.

We normally have our breakfast in the garden. Under the fig tree. Then we head to the Institute. To reach the train station we must go down two blocks, and then turn right, three more blocks, and there we are. Five minutes if I walk alone, less than three and a free workout if I am with Dr. Agnelotti. Quite fit he is, Dr. Agnelotti.

The way back is uphill from the train station to the Agnelotti's. This is usually when Dr. Agnelotti tells some nice story about his childhood, about how he met Mrs. Agnelotti, about their kids, about San Isidro. He usually tells the whole story without interruptions, while I try to continue breathing and to keep up with his pace.

San Isidro the hilly.

The Agnelotti family and San Isidro are so intimately related that it is hard to conceive one without the other. There could not be the Agnelotti without San Isidro, and there could not be San Isidro without the Agnelotti.

Yet, the essence of their relationship is freedom. San Isidro does not depend on the Agnelotti, and the Agnelotti certainly do not depend on San Isidro.

Where is San Isidro? Is it truly a Mediterranean village, or is it an expression of the character and the traditions of the Agnelotti? Is it located in Western Europe, which is from where the Agnelotti come, or is it located in the hearts of the Agnelotti, and therefore sitting on the Mediterranean because the Agnelotti incidentally live there? Who came first, San Isidro or the Agnelotti?

In the afternoon, when we walk from the train station to the Agnelotti's, we can smell different scents coming from everywhere. Those purple flowers over there, they smell really sweet. Right after the flowers, the marvelous smell of spices and herbs. What heavenly meal could ever have such smell...? Then the bakery, then the grocery, then we get home and of course straight to the kitchen.

San Isidro the perfumed.

Dips of smoked sausage, tomatoes, olive oil. The tomatoes come from the back garden. They are the pride of Dr. Agnelotti. No wonder. I had never seen tomatoes that size before, nor had I tasted tomatoes as delicious as those before.

The pleasure of cooking together. We all have our duties. I have been given the grand responsibility of peeling the vegetables, and then slicing the herbs and the leaf vegetables. A true recognition of my skills of a *chef*. If I behave well I can also do the cleaning.

After dinner, the ice cream shop.

More stories on the way to the ice cream and back. I speak less and less, and listen more and more each day. The stories are delightful. Every corner in San Isidro has something to tell, some anecdote, some love story, some past dispute, Gigino falling down from trees, Cicilla showing her paintings in the exhibit at the central square.

San Isidro the alive.

The town lives. It breathes, it smiles, it has its temper.

Dr. Agnelotti and his lovely wife are the interpreters. They read aloud the messages of San Isidro, so that I can understand them, despite being a foreigner.

After two weeks here, I can now venture to walk alone sometimes. I can wake up early and go to the bakery. I have already learned the tastes of Dr. Agnelotti and of Mrs. Agnelotti, so I can have the pleasure of returning their kindness – at least to a minimal extent – by sneaking to the bakery in the morning and buying some rolls for breakfast.

To my pleasure, I start to feel that sometimes I no longer need an interpreter. San Isidro speaks to me, and I can understand San Isidro.

San Isidro talks about many things. Merry and funny anecdotes, as well as heart breaking dramas.

San Isidro talks about all things and all people. San Isidro talks about everybody who lives in San Isidro. San Isidro talks about the whole of mankind. I follow with special interest what San Isidro has to tell me about the Agnelotti.

When I arrive back with the rolls, the table is already set for breakfast. Under the fig tree, as usual.

Of course I knew that figs came from fig trees. I had even seen a fig tree once. In a book. It had a trunk that went vertically from the ground up, full of branches and leaves. The leaves were green and stuck on the branches. Just like these ones.

Now I stand *under* the fig tree to have breakfast. There is a difference, I confess. Fig trees in books are not quite as fascinating as fig trees you can sit under.

The vegetables are planted over there. We can see them from here. The giant tomatoes come from there.

Gigino played football here when he was a kid. The Agnelotti's back garden was the place where all kids came to play. No wonder. There is something about this garden, some mystical magnetism that makes it the closest one could imagine of heaven on Earth.

Cicilla was always on her paintings. She used to stand over there, in the afternoons. Her paintings can be appreciated everywhere in the house of Dr. Agnelotti. There is also a beautiful landscape of a sunset in Sicily in Dr. Agnelotti's office, which was painted by Cicilla. That however was painted quite recently, I was told. Cicilla was already married when she painted that one.

The back garden of Dr. Agnelotti can change one's soul. No wonder the whole family is so creative. Each one on their own way, of course, and yet creativity and sensibility are present in all of them. Creativity and sensibility are present everywhere in Dr. Agnelotti's back garden.

The olive tree was planted by Cicilla. She was two years old when they moved to this house, and they planted the olive tree close to the front door soon after moving. It is amazing how slow olive trees grow. It is more than thirty years now since they moved to this house and the olive tree is just a little bit taller than me.

Unlike the fig tree. The fig tree is huge.

The leaves of the fig tree have a rough texture. They can be used to scour the cutlery, to great effect. If you want to have shiny spoons, plant a fig tree in your back garden.

The hennery is at the farthermost corner. There are reasons for that location, apart from the obvious fact that chickens are noisy and that a hennery naturally exhales a peculiar smell one would rather not sense from the bedroom. It seems to be the best place for the chickens, too, and also for the soil, so that the plants benefit from the manure.

This is from where the eggs come. Chickens, I mean.

As you may have guessed, I am a fine representative of the *homo urbanus* specimen. Yet, we are in an urban area. We are inside San Isidro.

Above that, however, we are inside the back garden of Dr. Agnelotti's home. A universe of its own.

Time flows differently in the back garden of Dr. Agnelotti.

A quiet place on the surface. If you stay here for less than fifteen minutes, I mean.

The door bell rings. It is Ms. Conchetta. She lives next door. Quite frequently Ms. Conchetta comes for a short chat and a cup of coffee by breakfast time. Ms. Conchetta is a very pleasant lady. She has lived alone since her husband died. Ms. Conchetta seems very healthy and strong, despite her old age.

Fifteen minutes later and Ms. Conchetta is off. In comes Mr. Fialho. Mr. Fialho shall work here for the day, cleaning the garden and looking after the tomatoes.

The door bell rings again. It is Mrs. Agnelotti's sister, on her way back after leaving her kids in the school. I should have brought more rolls.

Off goes Mrs. Agnelotti's sister, in comes one of Gigino's friends.

It is getting late, so now off we go, Dr. Agnelotti and me, to the train station. There is a fine full day of work awaiting us.

Good morning, San Isidro.

See you in the afternoon, back garden.

So many stories, so much life everywhere. San Isidro is the Agnelotti, the Agnelotti are their back garden. The back garden is the Agnelotti. San Isidro, the Agnelotti, the back garden. So different, so unique, and yet so inseparable.

It is hard to believe that the back garden is no longer going to be there four weeks from now.

The technical name for it is *director plan*. The municipality of San Isidro decided that a new road would be necessary to connect the central square with the highway. Someone decided that this new road would be important for the development of San Isidro.

Someone decided that the road should be built. Someone decided that the sooner it started to be built the better. Someone decided that the road should run through the back garden of Dr. Agnelotti's house.

The Agnelotti had gotten the warning six months ago. They would have to move. Of course the municipality of San Isidro would reimburse them for the inconvenience. They would have enough money to buy a nice new house. The back garden would have to be removed soon, but the house would take a little longer to be put down, so the Agnelotti would have time to find a good house.

The problem was, of course, that another house would not be *their* house. Another back garden, no matter how large and beautiful, would not be *their* back garden.

The fig tree would not survive. They were going to cut it down. The same with the olive tree, and with the tomatoes. The chickens would have to be sold.

And Giovanni is six months old.

Cicilla lives in a different town with her husband and little Giovanni. They visit San Isidro relatively often, but Giovanni is too young. He would not have the chance to know the Agnelotti back garden. He would not have the chance to walk around, smell the scent coming from the hennery, fetch gigantic tomatoes, and clean the garden.

Giovanni would not experience the history of the Agnelotti that was imprinted in that back garden. He would not have corners to see and remember memories of his grandparents, parents, uncle, neighbors, and so on. He would not see Gigino falling down trees when looking at that corner of the back garden, or his mother painting when looking at that other corner, or his grandmother singing and dancing during the summer storm when looking at the middle of the garden.

Giovanni would never sit down under the fig tree.

That seemed awkward to me.

That seemed outrageous to Mrs. Agnelotti.

That seemed unacceptable to Dr. Agnelotti.

The idea came one evening, at the ice cream savoring session. It sounded strange in the beginning, but after some discussion it started to sound reasonable.

Dr. Agnelotti would preserve the back garden.

Not physically, of course, since it was all set for the road to run through it. The back garden would be preserved somehow, so that little Giovanni could experience those memories and perceptions and stories.

Dr. Agnelotti was worldwide acknowledged as a top scientist, after all. There had to be some technological means to preserve back gardens for the future generations. There had to be some scientifically certified means to preserve the Agnelotti back garden for Giovanni.

This was not the initial plan when we set this thing up so that I could work with Dr. Agnelotti at the Institute. Yet, the idea seemed very challenging and scientifically exciting. So we changed our initial plans, and I began to work on the preservation of the Agnelotti back garden.

We are not working yet on how exactly we are going to preserve the memories of the back garden. Actually, we are at the moment on a preliminary step. We do not quite know *what* we are going to preserve.

What exactly is the Agnelotti back garden?

Indisputably, the Agnelotti back garden is a garden. As such, it inherits the properties of all locations on Earth that can be classified as gardens.

That could be a good start. Gardens are clearly bounded locations, as well as for example swimming pools, airports and shopping malls. Clearly bounded locations, in turn, belong to the more general class of places on Earth.

All we have to do, therefore, is to find a good starting point, i.e. a generic category with clearly defined properties, and derive from this generic category more and more specific categories, by adding new features and properties and values to variables, until we get to the essence of the Agnelotti back garden. With that, Giovanni will have a very refined notion of the garden, which is also going to be a precise scientific model of the garden. Giovanni will also have a lot of additional information about more generic categories, which can come out to be useful when he gets to school age.

The problem is that this approach may amount to a rather bulky collection of data for young Giovanni.

We may need to select some relevant aspects of the back garden – as well as to the generic categories to which it is a particular instance – to include in our description, and to leave out some other aspects.

We have considered that, given the physicality of the Agnelotti back garden, aspects that characterize its physical properties can be a good choice.

We have started with the physical properties that can be perceived by the five primary senses. Given the limited time and budget we have for this project, we may need to limit the scope for the main senses, so we decided on a list of priority of the five senses.

That would be for BackGarden version 1.0. Later versions can of course include the remaining senses.

We have started with vision. The Agnelotti back garden must be carefully measured, and a detailed measuring may take a little while, considering that the walls that separate it from the streets are not very straight.

These are ancient walls. They have been here for at least ninety years. Especially that one at the back. The bricks are beautiful. They were made the traditional way. They are not very regular, but they are very resistant.

Let us get down to the objectivity of facts, shall we?

We can of course approximate the shape of the walls to a polygon. It has four sides, none of which parallel to any other. We can come later with a professional tape line and write down all measures.

The fig tree is not close to the walls. It is not quite in the middle of the garden, also. It stays closer to the wall that faces the alley, farther from the house. I would say two thirds from the house, one third from the wall that faces the alley.

That would be like six meters from the house and three meters from the wall of the alley. As a simplification, of course, so that we have round numbers.

Not considering the diameter of the trunk.

About six meters from the house and three meters from the wall, to the geometric axis of the fig tree.

Assuming that the fig tree is a perfect cylinder, standing perpendicularly from the floor.

The floor is relatively even. Not perfectly horizontal, but we can assume it is as a first reasonable approximation. That makes the fig tree a perfect cylinder, perfectly vertical and standing on the perfectly horizontal floor on which stand the Agnelotti's residence. Six meters minus its radius away from the perfectly smooth wall of the house, three meters minus its radius away from the perfect wall that separates the perfect garden from the perfect alley.

The hennery is at the northeast corner of the garden. It holds five chickens and a rooster at the moment. They lay eggs, which are conventionally egg shaped. The chickens, I mean. The hennery is about three square meters large. Pretty comfortable for five chickens and a rooster.

Mrs. Agnelotti knows each chicken well. It is hard to believe how well the chickens – and the rooster, of course – seem to know Mrs. Agnelotti as well. They have names, I have been trying to memorize their names but so far I did not manage.

The tomatoes are at the northwest corner of the garden. It may be relevant to record their dimensions. They vary, of course, so we may estimate an average.

It will be nice to record the taste of the tomatoes, too. I do not know how we will make it, but for the moment we are focusing on vision anyway. We can leave this for later consideration.

The tomatoes will have to be removed, anyway. So there is no problem if we eat more tomatoes than we used to from now till the day the tractors will come.

Incredible tomatoes, these ones. I did not know tomatoes could be so big. I could never imagine tomatoes could taste so good.

Dr. Agnelotti's father is the one to blame. It was his idea to start planting the tomatoes. He used to come here every day to water the tomatoes and to clean the ground around them. Apparently he also put some combination of compost, minerals and herbs on the ground once a month. After he grew older and this became too heavy for him, he passed the responsibility to Mrs. Agnelotti. She has evidently learned the deed.

Mr. Fialho helps a lot, especially with the heavy work.

Twenty five centimeters. Plus or minus two centimeters. This is the estimated average diameter of the tomatoes. Giovanni will know that, apart from exceptional cases, the tomatoes at the Agnelotti back garden measured between twenty three and twenty seven centimeters.

Visual perceptions can be very rich, and we are well aware that it may take some effort and some time to record sufficient visually perceptible characteristics of the back garden so that Giovanni can appreciate it even after it is no longer here.

Visual perceptions are however not sufficiently rich to convey the necessary perceptions so that Giovanni can have the experience of being in the back garden.

Considering this statement as a provisional assumption, and again taking into account our limited resources and time to develop this work, we have decided to parallelize a little our activities, and work with hearing together with vision.

Sounds are a little more complicated to work with than geometry. Apart from very specific exceptions, sounds are not emitted constantly and regularly all the time.

Certain sounds occur with approximate regularity every day. As a first approach, we are focusing on these sounds, and leaving the incidental sounds for later consideration.

Six o'clock.

Sparrows. They come every morning. They sing loudly and merrily. They are certainly a necessary detail to be remembered. Little creatures, so fragile and yet so strong.

I am certainly not the only one to like the fig tree.

The sparrows came first. Their relationship with the fig tree is older and far more intimate than mine.

Six o'clock.

Sparrows.

There are usually between five and twelve sparrows. You would imagine there are at least fifty, by the noise.

Seven o'clock.

Ms. Conchetta.

A short ring at the door bell. Then a cheerful "Good morning everybody". Then the usual morning chats, sipping a cup of coffee down the fig tree.

A quarter past seven.

Ms. Conchetta speaks "Goodbye folks, see you all later" and leaves. Mr. Fialho comes in. "Good morning". The door of the workshop whines open. Mr. Fialho gets in to get the tools. Clanks and bumps. Mr. Fialho gets out. The door whines shut.

Forty past seven.

The door bell rings. A cheerful and loud “Hello”. Laughs and kisses. Mrs. Agnelotti’s sister. Chatter. Coffee cups being drunk.

Sounds and images must then be combined. Here we go.

Seven o’clock. The door bell rings. Ms. Conchetta comes in. A short old lady, about one and a half meters tall. Colorful clothes. A pleasant smile under thick glasses. A loud and cheerful “Good morning everybody”. The sky is blue. The fig tree is very green. At least at this time of the year.

Poor Giovanni will have a lot to read. Or listen. Or sense by whatever appropriate means the information that is going to be recorded for him.

The appropriate media will have to be employed. Can written text be good enough? Do we need photographs? Video? Recorded sound? How much do we need of sound fidelity? Do we need some sort of three dimensional rendering?

What is really the point, after all? What do we want to preserve for little Giovanni?

Our routine question. Day in, day out, here it is.

What *is* the Agnelotti back garden? Where does it start, and where does it end?

How many bits of information do we need to capture the true essence of the Agnelotti back garden?

What about the bits that are decided to be left out of the characterization of the Agnelotti back garden?

What are these bits?

Which are these bits?

How many bits are we going to need to start talking about Gigino and his fixation to fall down trees? About Cicilla painting in that corner of the garden?

The average height of Gigino's falls can be estimated to be around two meters. The problem of course in this case was the standard deviation. It was quite high. With that little Giovanni would be able to infer that Gigino did fall from as high as four meters, thus explaining Dr. Agnelotti tachycardic attacks.

How many bits are going to be needed to convey the sense of Mrs. Agnelotti singing in the summer storm? Not the song, not the sounds, but the true sense of those moments?

How relevant is the diameter of a tomato? On the other hand, how to inform little Giovanni about the tomatoes? Would a photo be helpful? Perhaps a mummified tomato in a bottle?

What *is* the Agnelotti back garden? Where does it start, and where does it end?

Chapter 3

Information Flow

One starts off writing with a certain zest, but a time comes when the pen merely grates in dusty ink, and not a drop of life flows, and life is all outside, outside the window, outside oneself, and it seems that never more can one escape into a page one is writing, open out another world, leap the gap. Maybe it is better so.

Italo Calvino

Now that we have the notion of information systems as representations of observable structures found in the reality, we develop the idea that information systems can be communicated between agents.

A point to be considered is that the information system transmitted by an agent is not the same one perceived by that agent. As discussed in the previous chapter, information systems are, so to speak, native to the reality. A communicated information system is the result of the action of an agent, who is assumed therefore to have selected what to communicate to other agents. Agents behave as mediators between information systems and other agents in the communication process, and this mediation interferes with the original structure and contents of the information system to be transmitted. Hence, an agent perceives an information system, interprets it, builds a second one and transmits this second information system to other agents.

As a consequence, a communicated information system brings a second layer of separation between an agent and the reality. The communicated information system carries with it some selected information from an original information system, together with goals and intentions of the sender agent, expressed in at least three different forms:

1. How and why the agent selected the original information system in the first place.
2. What the agent selects to communicate to other agents, and how it organizes its messages.
3. How the agent communicates its messages.

For the receiving agent, a different reality presents itself for appreciation, which includes the communicated information system and other pieces of information, including the interaction between the receiving agent and the world, the interaction between the sender agent and the receiving agent, and possibly the interaction between the sender agent and the world. The *raison d'etre* of the mediation performed by the sender agent is to filter specific aspects of the reality to be presented to the receiving agent, so the expected situation is that in which the receiving agent does not interact directly with the reality as presented to the sender agent.

In the following sections we discuss these points in greater detail.

Communicated information

The construction of a message to be communicated is an action performed by an intelligent agent. The result of this action is an information system, i.e. a structure of pieces of information. It can be correlated to another information system perceived by this agent, for example when it describes an event that has been observed by the agent, but not necessarily so – the agent can be a novelist, for example, and the information system it builds can be a purely fictional story.

In any case, this constructed information system is correlated to a reality that includes the agent that has formulated it. It reflects some aspects of that agent, how it sees the world, its plans and intentions, and what it has selected to communicate to other agents. It does not refer to *the world* with which the agent interacts, rather it refers to *a world* inhabited by an agent who has formulated its view about the world and is communicating this view to whoever is out there to receive this message. Differently from the information systems considered in the previous chapter, the information system to be communicated is built by an agent. It is an artifact, made of a collection of abstract symbols embodied in some sort of media: sounds in the case of spoken language, written symbols in a piece of paper, a piece of software running in a computer, etc. As any information system, it is incomplete and may have limited accuracy.

The language and the media chosen to represent an information system impose certain limitations on the completeness and accuracy of the system. For example, if the language being used admits only a finite amount of sentences, the facets of the world that can be captured by this system will also be limited to a finite collection of aspects. Some facets of the world will have to be left behind, and perhaps some of the facets being considered in the information system may need to be joined together in equivalence classes. The completeness of an information system is therefore limited by the language employed to encode it.

On the other hand, if the language being used to represent an information system presents ambiguity, the facets of the world that can be captured by the information system cannot be sharply distinguished. The same facets of the world, when represented for example using mathematical logic, technical English or colloquial English may result in information systems with different degrees of precision.

The information system to be communicated is, therefore, a simplification of the reality as perceived by the agent who formulates it, based on an information system – called the "original" information system above – which refers to a world which includes the agent itself. The usual name adopted for this simplification is a *model*. We assume, therefore, that *model* is a synonym of *artificial information system*, which is going to be the term we preferably are going to employ.

Noise and distortion

What the receiving agent gets is an information system whose corresponding reality contains the sender agent and an *artificial information system* specially built by the sender agent to be communicated to the receiving agent (Figure 3.1). In principle, the artificial information system correlates with the information system perceived by the sender agent. This artificial information system should ideally be *perfect*, i.e. complete and precise enough to have a one-to-one correspondence with the original information system to which it relates.

With a perfect artificial information system, the receiving agent would get an enriched perception of the reality if compared with what had been captured by the sender agent, in a specific sense. Namely, the receiving agent would have access to precisely the same as what had been captured from the reality by the sender agent, plus additional information related to the sender agent itself (which for example can inform about goals, intentions and other specific features of the sender agent – see Figure 3.2) and to the opinions of the sender agent about the receiving agent.

The typical situation, however, is to have imperfect artificial information systems being built by the sender agent. The sender agent, therefore, distorts its original perceptions when building an information system to be communicated. The distortion is influenced by the sender's goals, intentions and specific features, and takes into account the receiving agent(s) to whom the artificial information system shall be communicated.

One possible way to depict all this is the following: the potential receiving agents who can be the object of a communicative act from the sender agent are part of the reality perceived by that agent, from which an information system will be selected and/or characterized. The sender agent itself also perceives itself as part of this same reality, of course, thus taking into account some of its goals, intentions and specific features in the information system it perceives. Based on all that, the sender agents build an artificial information system, somehow related to the original one that has been perceived by the sender agent.

A receiving agent, on its turn, perceives a reality comprised by the sender agent, the artificial information system built by that agent, and of course the receiving agent itself, thus taking into account some of its goals, intentions and specific features. That is the receiving agent's reality, from which a new information system will be selected and/or characterized.

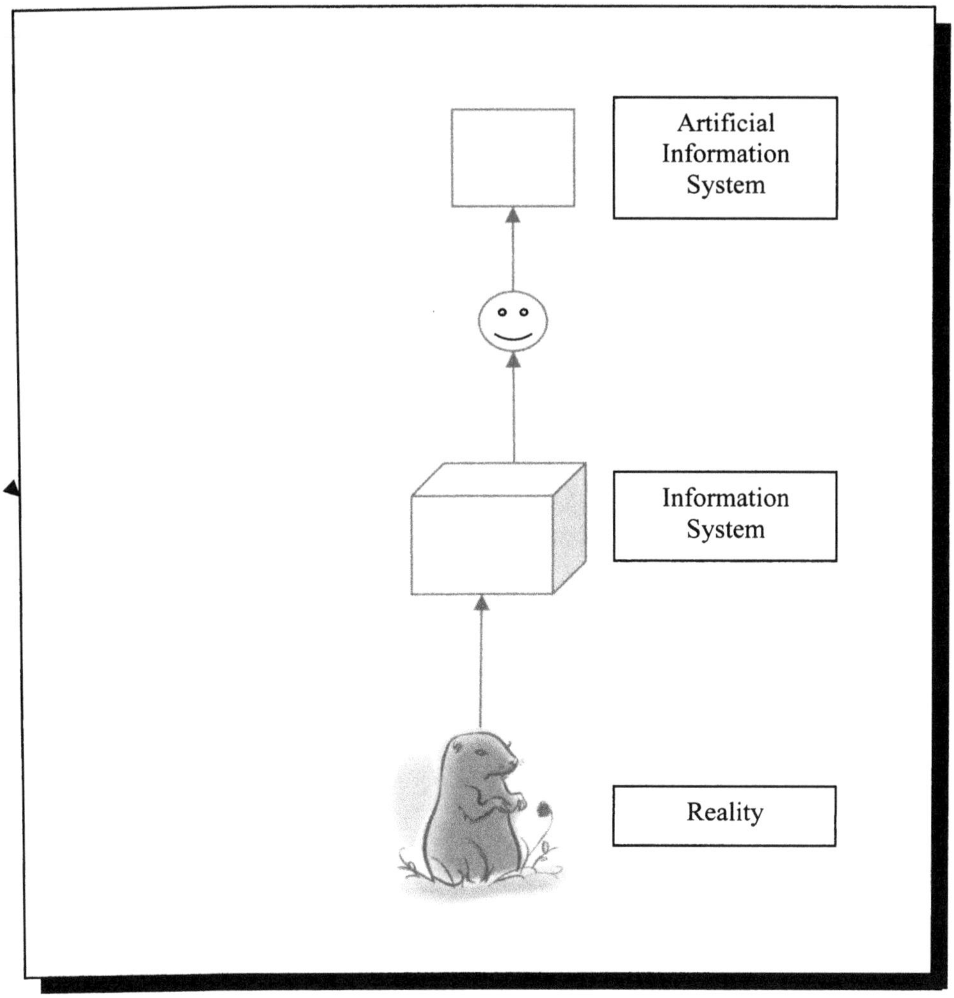

Figure 3.1. Artificial information systems

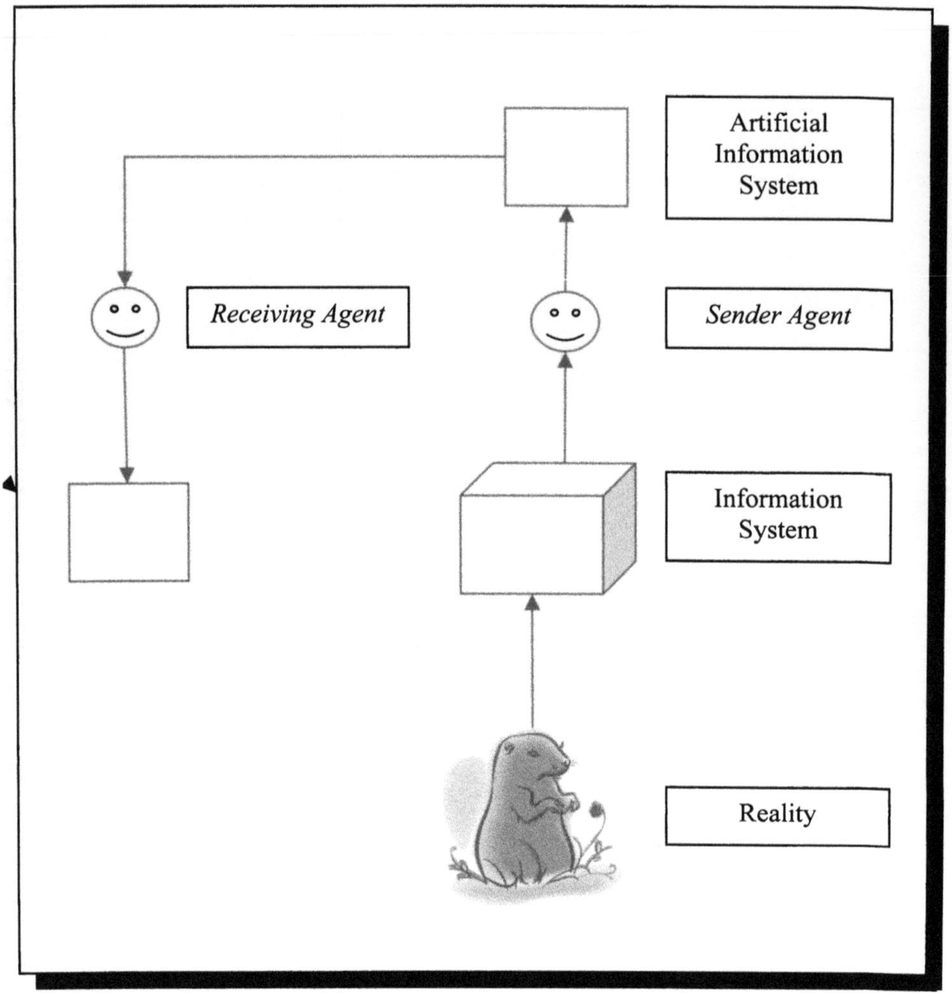

Figure 3.2. Perfect communication using a perfect artificial information system

To give a concrete example, we – Flávio and Jaume, the authors of this book – have little hope that you – dear reader – will share with us, after reading the book, the *perceptions* and insights that have taken us to write about information flow and knowledge sharing. At its best, this book shall reflect *some aspects* of our perceptions and insights – those that we have decided to name Information Flow and Knowledge Sharing (i.e. the title of the book) – already distorted by our goals, intentions, beliefs, prejudices, etc. – which by the way are also expressed – at least partially – in this very same book.

When writing the book, we imagined a prototypical reader, to whom we wrote. This prototypical reader is part of the reality we perceived while formulating our views about information and knowledge. Among our goals, we highlight:

1. The interest in shedding some light on issues we consider important and useful for information system architects and information system managers, as well as for those professionals who are usually named "knowledge engineers" and "knowledge managers"; and

2. The interest in bringing to discussion some issues related to our social and economic organization, the underlying beliefs that drive the construction of this organization, and some possible future developments from the point where we stand at present.

The book is an artificial information system. Its quality depends on how well we managed to capture in it relevant features of:

1. The original reality we have perceived, which includes structures that can be described using the concepts we have described – information, reality, information systems, etc.;

2. Our own goals, intentions and specific features; and

3. The relevant features of the potential readers to whom we have addressed these lines.

Its *utility* naturally depends on other things, e.g. our talent to consider as relevant those features that are indeed relevant for the largest possible group of readers, which shall include the majority of information system architects, information system managers, "knowledge engineers" and "knowledge managers".

Books are, therefore, fine examples of the artificial information systems referred to above. So far, however, we have considered the communication process itself as perfect and limitless. Let us imagine, however, that in order to save paper and make lighter books, an eccentric publisher decides to delete all vowels from a text. Let is imagine, moreover, that in the course of deleting the vowels the publisher commits some mistakes and misplaces some of the consonants.

It is the book with no vowels and containing mistaken consonants that will be perceived by the receiving agent, and thus the noise and distortion inserted in the process by the communication channel is a point not to be neglected when analyzing information systems and how they flow between agents.

If a writer is aware of the eccentricities of the publisher, she may change her writing style and choose words that can be identified even with the vowels deleted. Likewise, the more a reader knows about the imperfections of his book that originate from the publishing process, the better he can get by with them.

Different communication media give rise to different sorts of imperfections to be taken into account: books have misspellings and missing pages, telephone calls have background noise and echo, and computer programs have bugs (Figure 3.3).

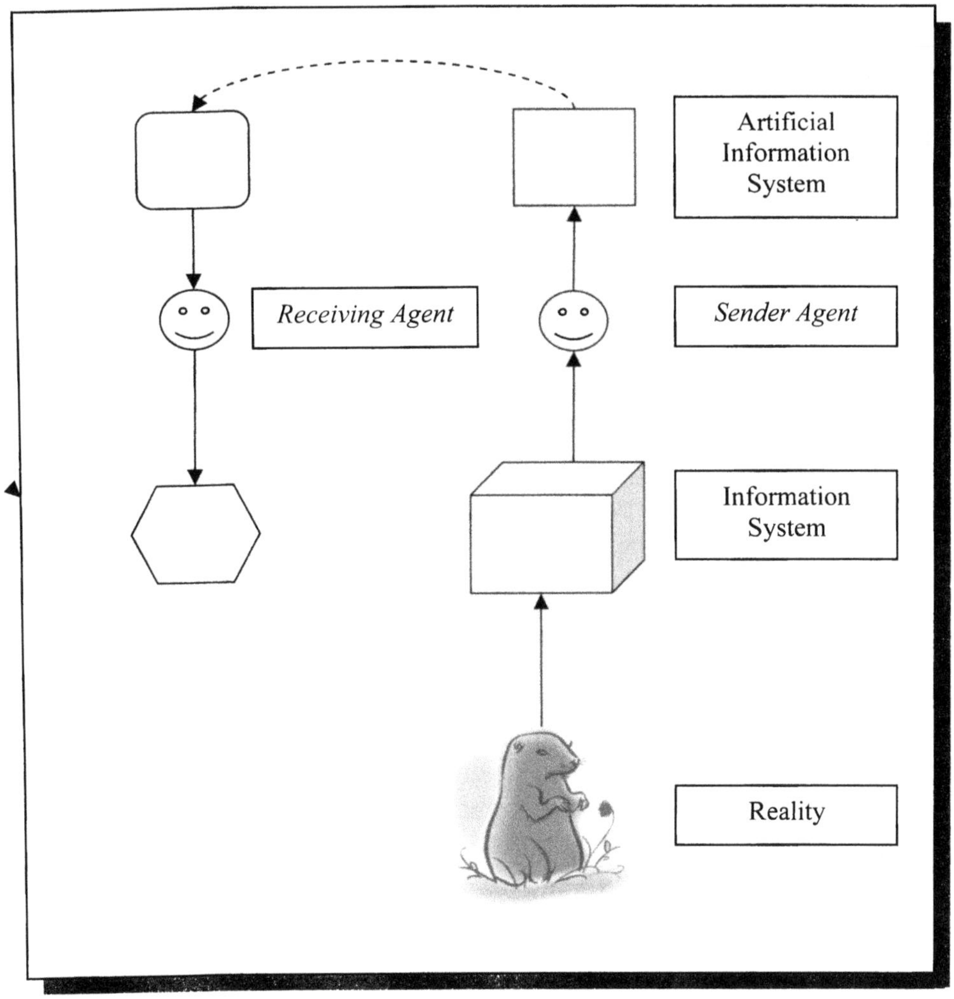

Figure 3.3. Imperfect communication using an imperfect artificial information system

Received information

This is a good point to remind that there is no difference between the receiving agent and the sender agent. They are both *agents*, who provisionally assume different roles in a communication process, thus enacting on a particular way of interaction. During this process, the sender agent and its communicated artificial information system become part of the reality perceived by the receiving agent, together with the noise and distortion inherent to the communication process.

Typically, the receiving agent does not have direct access to the same reality as that perceived by the sender agent. A curious recurrence relation is worth observing, if we consider a chain of agents A_1, A_2, etc., such that agent A_i acts as the receiving agent for agent $A_{(i-1)}$ and as the sender agent for agent $A_{(i+1)}$ (Figure 3.4). Any of us takes part in chains like this most of the time in our everyday life. To give a simple example, when watching the news on television we act as receiving agents to the person who is presenting the news, who in turn is probably reading a text prepared by someone else, and therefore is the sender agent to us and the receiving agent for another agent, who may be the field reporter who has had the live experience of the news being presented, and so on. We may later comment on the news we have watched, thus becoming senders and this way continuing the chain.

When we look at this recurrence relation backwards, we find ourselves in the middle of the chain or at its right endpoint when we receive some information and keep it for further use or transmission. We may ask ourselves "where does a chain start?" In other words, who is agent A_1 in any chain of communications in which we participate? Do we ever have direct access to reality, or are we permanently setting ourselves apart from it, thus confining ourselves solely to mediated relations already filtered by goals, plans and intentions of other agents, as well as ours?

When we have specific goals and intentions, our perceptions are driven by them. We do not see all of what is presented do us, but what we choose to see from the host of existing possibilities, actively blinding ourselves from a myriad of possibilities in order to be "practical" and reach efficiently our goals, fulfill our intentions, and so on.

Our goals and intentions also drive the construction of our artificial information systems, which are then communicated taking into account the receiving agents of our messages and the communication media.

When we adopt the stance of receiving agents, instead of sender agents, the reality that is presented to us is the reality of artificial information systems, sender agents, communication media, and glimpses of the world that not necessarily coincide with those glimpses perceived by the sender agents. This reality can also be broken into a multitude of information systems, among which we choose some based on our own actual goals and intentions. These selected information systems shall form the basis of the artificial information systems that we are going to communicate to other agents, and so on and so forth thus forming the communication chain referred to above.

Communication requires reliability and ultimately trust. When two agents engage in communication, some portion of information shall go unchecked – otherwise the receiving agent would be accessing directly the reality perceived by the sending agent. The receiving agent assumes that the received information is reliable, that the distortions inherent to it are acceptable and that therefore it can give up direct connection with the reality that triggered the construction of the received information systems on behalf of the communicated information.

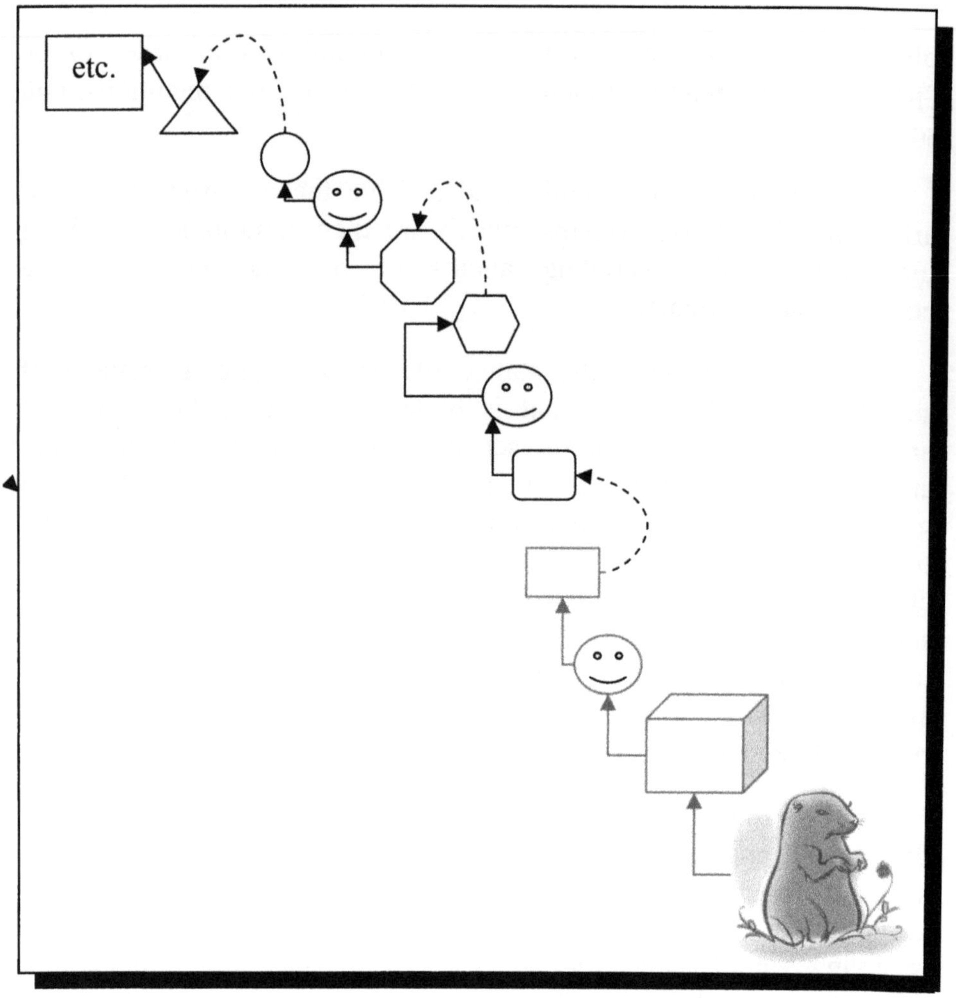

Figure 3.4. Communication chains

Communication systems

A *communication system* is a selected medium through which agents engage into communication, i.e. exchange information (Figure 3.5). Essentially, the communication system must provide the means for the sender agent to encode its artificial information system, and for the receiving agent to read this artificial information system. Ideally, the receiving agent should be able to read the artificial information system in its entirety through the communication system, although this must not be necessarily the case.

As mentioned before, the communication system imposes on the communication process constraints, inherent to the nature of the selected medium to encode information systems. For example, if the sender agent decides to interact with the receiving agent through written media – for example, by writing books – then some aspects of any perceived reality can be transmitted, but others simply cannot (sounds and voice intonations, for example, can be described in a text but not really *represented* in the text).

Consequently, depending on what is intended to be communicated, certain media are more appropriate than others. The decision of what media to employ can be complex and depends on many attributes, among which we highlight the goals and intentions of the sender agent, and the perceived information system that this agent has at hand – i.e. how reality has influenced and informed the sender agent – including in this reality the way the sender agent sees itself and the way it sees the potential receiving agent with whom it will communicate.

As for the complexity of the artificial information system built by the sender agent, the sky is the limit. A written message can range from a note stating that the agent is out for lunch to James Joyce's *Finnegans Wake*; a piece of music can range from a nursery rhyme to Beethoven's *Große Fuge*; and so on.

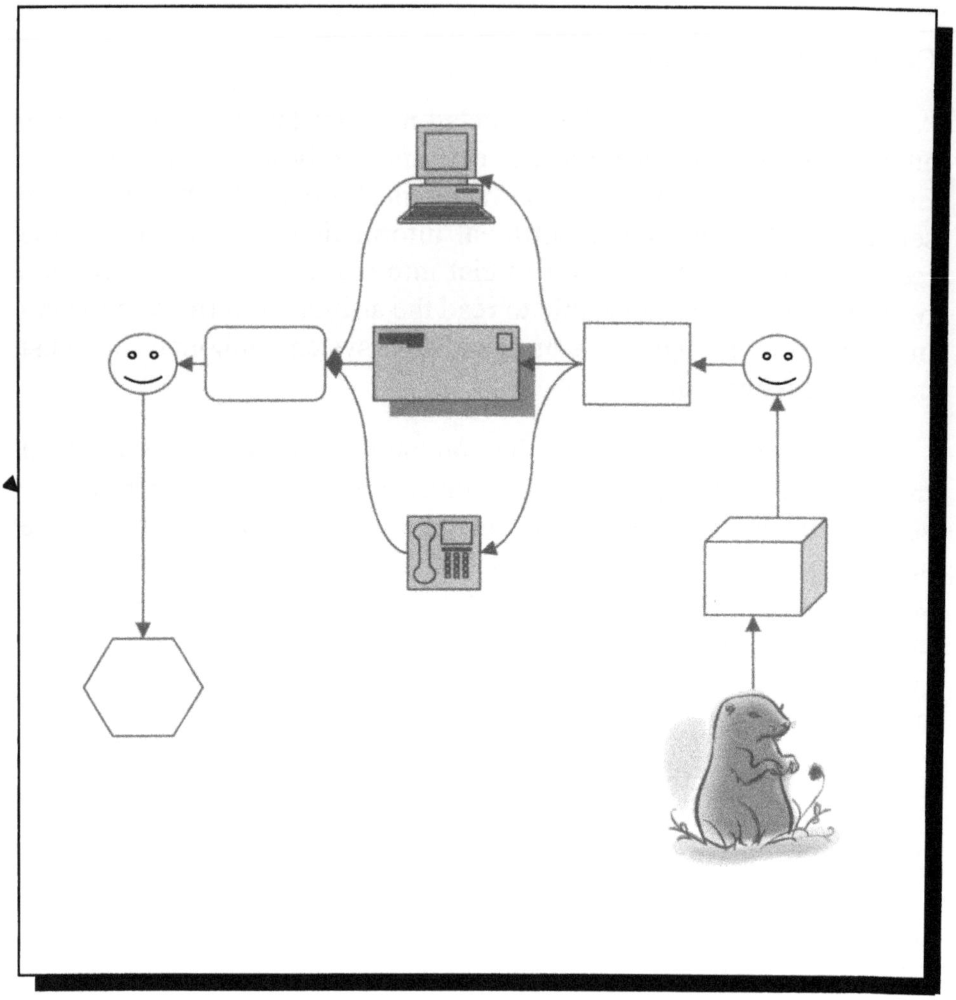

Figure 3.5. Communication systems

In this work we envisage computer programs as artificial information systems, through which two or more agents communicate. This approach to conceptualize computer programs leads to certain design values and criteria, which are not necessarily universally adopted. This is the approach we have employed and which we propose to drive the design of computer programs and systems.

In some cases it is fairly natural to view computer programs as communication devices. Take an instant messaging system, for example, which has been designed for the primary purpose of enacting communication between people. Even in such cases, however, we emphasize that *three* agents are participating in the conversation, instead of just the obvious two users who entangle in a dialogue through the program: the two users plus the programmer who deployed the system. Indeed, to analyze the use of an instant messaging system based on our view we must do it in two steps:

1. In a more foundational level, the programmer is the sender agent and the two users who entangle in a dialogue are receiving agents. The reality perceived by the programmer includes the way people interact, from which an information system is captured that corresponds to a particular type of interaction named *dialogue*. The programmer devises a model of dialogue, based on her perceived notion of a dialogue, her goals and intentions and a model of her receiving agents, namely the potential users of the messaging system. An artificial information system is then built, through which interactions will occur limited by the artificial information system, the goals and intentions of the programmer and the expected behavior of the users implicit in the model devised by the programmer.

2. In a second level, two users employ the instant messaging system to engage into conversation. In order for this to happen, the users must first perceive the reality that is presented to both of them, which includes the artificial information system proposed by the programmer – i.e. the instant messaging system *per se*

> – and also the goals and intentions of the programmer and the proposed model of users of that system, which the users must accept. The presented reality also precludes momentarily other sorts of interaction, this being the whole point of having and using an instant messaging system to interact. The users then take turns as sender and receiving agent during the dialogue, not any more in their initial realm of human interactions but now on this virtual world proposed by the programmer.

It is as if the programmer were a scriptwriter, preparing a play in which some improvisational performance is accepted. The users are the actors, and yes, it is the scriptwriter who selects the qualities an actor should have, and not the opposite. In other words, the programmer chooses, when preparing his message – *aka* software system (in our case an instant messaging system) – the people who are going to use it. To the potential users – *aka* receiving agents – have the possibility to accept the role in order to engage into acting in the virtual world proposed by the programmer.

Let us consider now a multi-user computer game. From the standpoint suggested in the previous analysis, there is not much difference between a game like this and an instant messaging system. In both cases, the users assume specific roles proposed by a programmer in order to interact according to a protocol specified in an artificial information system, and by doing so agree to participate in a virtual world. The major difference between the first and the second example is that in the second case the users deliberately accept a world that is being proposed by a programmer and accept to momentarily give up their "real lives" to dive into the virtual world.

Let us continue to move progressively on through a series of examples, to illustrate the point that explains why we have used quotes around "real lives" in the previous paragraph. If we consider an action computer game instead of a multi-user game, it can be the case that a single user interacts in a virtual world with synthetic characters whose behavior has been programmed in advance. No matter how sophisticated these synthetic characters are they belong to the artificial information system proposed by the programmer – the sender agent – as part of a message being communicated to the receiving agent – the player. They are, in a sense, a sentence in a dialogue occurring between the programmer and the player.

Consider now a multi-agent system in which a collection of intelligent agents interact to solve a complex engineering problem. One possible way to understand this system is: when faced with a complex problem to solve, a programmer once built a model in which she expressed her understanding of that problem and how it could be solved. This model was built considering the problem itself and a proposed solution, and also specific goals and intentions that the programmer had and a prototypical engineer who could one day want to solve the same problem. When an engineer tries to solve a problem, he checks whether his problem can adjust to the model previously proposed by the programmer. If this is the case, the engineer accepts the role specified in the definition of the prototypical engineer, and adjusts his problem to the proposed model. This way, he can communicate with the programmer.

We do not preconize that the only way computer programs should be interpreted is as messages being exchanged between programmers and users. We do not even claim that this is a good way to interpret all computer programs. We suggest that every computer program *can* be interpreted this way, and that looking at computer programs this way can unveil features – desired, undesired or unexpected – that a specific program can have, thus being an interest tool to design good programs.

Moreover, our view is a call to remembering that, after all, the fundamental reason to build computer programs – or any other technological artifact – is to propose new forms for people to communicate with each other, in more or less controlled ways (Figure 3.6).

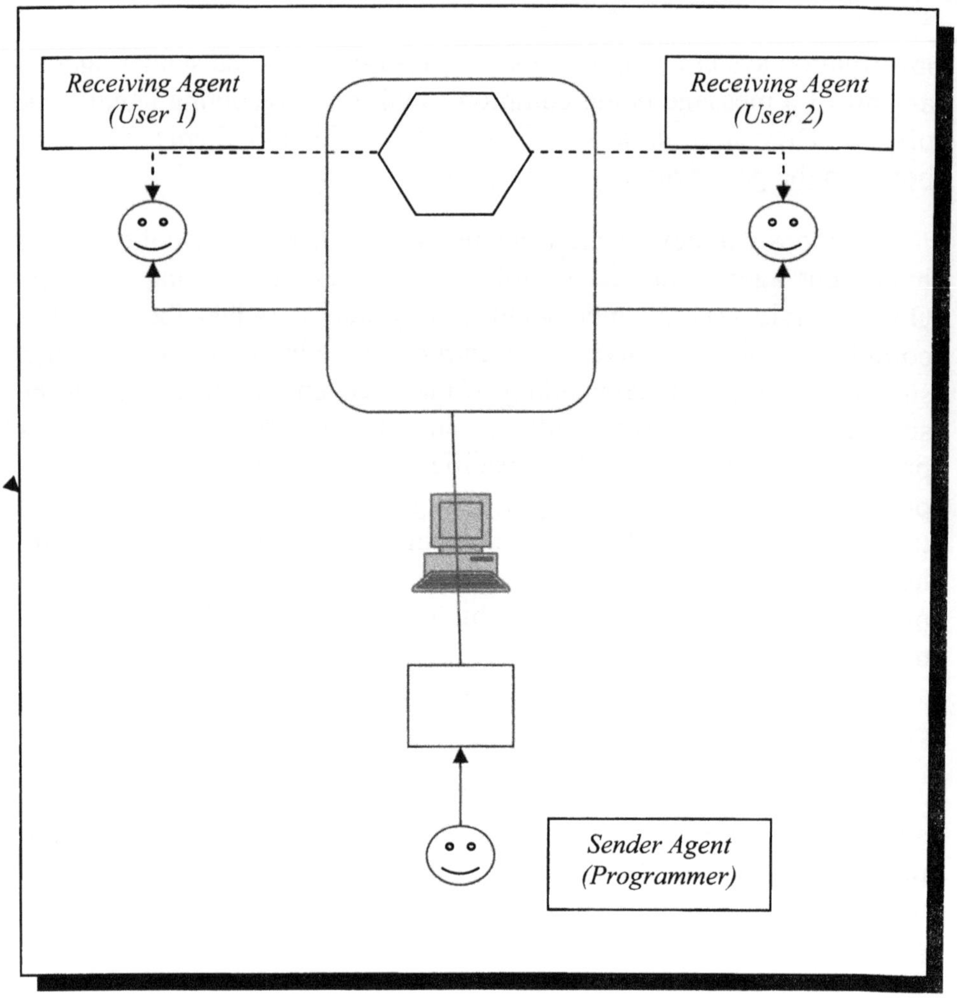

Figure 3.6. Computer programs as communication systems between programmers and users, and among the users

The purpose of communication

We identify three reasons for agents to communicate with each other. These reasons are typically invoked by other reasons – for example, a violinist is the receiving agent for a composer, and engages in this communication process in order to play. Playing music is the invoking reason, which brings about one of the three reasons to communicate as detailed below. The reasons for agents to communicate with each other are:

1. *Communication to explain:* an agent wishes to understand some aspect of the reality. For some reason, the agent cannot access that aspect of the reality directly – for example, it may be geographically or historically distant from the agent, or it may be too costly or time demanding for the agent to connect directly with that aspect of the reality. The agent then accepts the position of a receiving agent to a sender agent who has the means to access reality – for example a traveler who has been to distant places, a historian who has studied thoroughly past events, or an expert in a field who is prepared to share his notions of that field through lectures and textbooks. Receiving agent and sender agent engage into a communication process, in which the receiving agent replaces his interest in directly connecting with reality by the connection with the artificial information system proposed by the sender agent, together with all presuppositions these agents will have about each other and about themselves.

2. *Communication to command:* this is the case of the violinist referred to above. An agent wishes to *build* something, i.e. interfere directly upon reality through action. This agent has a goal to reach and looks for instructions to act in order to reach its goal the most effective way. These instructions can come through communication with a second agent who has been

through a similar situation wishing to reach a similar goal, and who has been capable of recording the procedures it has followed as an artificial information system. Alternatively, the second agent may be an expert in the field of interest of the first agent – i.e. an agent who has had the opportunity to access directly the relevant portion of reality for the goal to be reached in greater detail than the first agent. Either way, the second agent – a sender agent – provides to the first agent – a receiving agent – an artificial information system conveying instructions for action. The success of the action of the receiving agent – which can be measured in terms of how well its goal is reached – is going to depend, among other things, on how well the reality of interest for the receiving agent matches with the one conveyed in the artificial information system proposed by the sender agent, as well as how precise are the mutual presuppositions of the sender and the receiving agent. Typical examples of this type of communication are:

a. A violinist wishes to convey certain emotional content. In order to do so, she looks for the appropriate music score to perform. The composer is a sender agent, the music score is the artificial information system and the violinist is the receiving agent in this case.

b. A construction engineer wishes to build a house to bring comfort to a specific family. In order to do so, he looks for the appropriate blueprints on which to work. In this case, the architect who prepares the blueprints is a sender agent, the blueprints are the artificial information system and the construction engineer is the receiving agent.

c. A programmer wishes to write a program to fulfill specific requirements. In order to do so, she looks for the appropriate software specification. The system designer who prepares the specification is a sender agent, the specification document is an artificial information system and the programmer is the receiving agent in this case.

3. *Communication to satisfy:* an agent may wish to relinquish its participation in reality on behalf of a proposed reality conveyed by an artificial information system. In this case, it is the desire of the agent to shut down the connection with reality, and replace it by a virtual reality of some sort. The receiving agent, in this case, looks for an appropriate artificial information system – prepared by a sender agent who has an appropriate understanding of the receiving agent, and whose nature is also sufficiently known and accepted by the receiving agent – to adopt as a virtual reality to inhabit. An obvious example is a virtual world like Second Life®. More subtle examples are an instant messaging system as mentioned before, or any public space within which individuals accept roles and behavior according to strictly defined protocols, like for example an airport for the traveler wishing to check-in for a flight or the branch of a bank.

Bibliographical notes

Communication restrictions on the flow of information imposed by the communication media were explicitly and rigorously discussed in (Dretske, 1981).

Our proposed view of software systems as artifacts to enable interpersonal communication, including communication between users and designers, is aligned with the notion of *interaction design*, advanced by Winograd and colleagues in (Winograd and Flores, 1986; Winograd, 1997). A fine publication devoted completely to interaction design is (Sharp, Rogers and Preece, 2007).

Trust is a notion whose importance has grown substantially in recent years, in diverse areas such as the design of autonomous systems in artificial intelligence, management theory and economics. An interesting book on trust is (Solomon and Flores, 2003). Trust and reliability, however, are very different concepts – as discussed for example in (Solomon and Flores, 2003), and in the chapters to follow – although they are both important for the exchange of information systems. These two concepts should not be treated as synonyms.

The reasons for agents to communicate we have depicted in this chapter are directly related to and inspired by (Borgmann, 1999). Our notions of communication to explain, communication to command and communication to satisfy correspond to same extent to Borgmann's notions of information *about* reality, information *for* reality and information *as* reality. Essentially, we have endeavored to bring Borgmann's view to within our setting, to enrich and illuminate our conceptualization of information.

Intermission

We have done our best to withhold to the scientific objectivity of facts, in order to preserve the essence of the Agnelotti back garden with the utmost fidelity.

It has proven to be a great challenge, though, to determine the appropriate criteria to identify the information that should be preserved about the back garden.

In order to determine these criteria, we started to discuss what criteria should be used to identify the appropriate criteria to preserve the essence of a back garden – *any* back garden, for that matter – which soon led us to discuss whether there could be any criteria to select the criteria that should be used to identify the appropriate criteria to preserve the essence of a back garden. We started to suspect that this discussion would not end, unless we moved to a meta discussion about the nature of the selection of criteria – *any* criteria, for the selection of whatever could have to be selected – which is clearly a recursive problem.

Usually, in mathematics, recursive problems can be dealt with by means of inductive theorem proving. We started considering that maybe some transfinite induction techniques could be helpful, given the apparent structure of the entanglements found in the formal characterization of the problem we then started to call the "criteria selection problem", or *CSP* for short.

We then decided to leave aside for a moment the back garden *per se*, to study for a while this interesting problem in metamathematics that we were starting to characterize. We could perhaps publish a couple of articles on that issue, after proving some results. It could take a while, however, since we would first have to find the appropriate language to pose the problem, such that an algebraization of the meta level reasoning could be stated with some elegance and simplicity.

That of course would naturally lead to the philosophical issue of what a base case for this inductive process would be like, which could provide us with discussion material for a couple of other articles on philosophy and methodology of science, as well as another couple of articles on realism, materiality and materialism in natural sciences as well as in epistemological phenomena in general.

Another interesting spin off of this problem could relate to the computational complexity of the *CSP* – or at least of the decidable fragments of this problem. We spent one afternoon discussing whether the decidable fragments of the *CSP* could be separated from the undecidable fragments. We wrote down a few notes on the decidability of the separability of the decidable fragments of the *CSP*, which was evidently a metametamathematical problem.

Then there was of course the issue of implementing special purpose automated theorem provers to deal with the decidable fragments of the *CSP*. This would be a nice problem indeed to work on. We could try to find an interested PhD student to write the code, and then build some computational experiments related to the *CSP*.

It was good that we wrote down all these ideas in our notebooks. We must address all these interesting problems, or perhaps suggest some of them as projects for our students, lest they are lost in the middle of many other unfinished ideas and scribbles that we have collected along our many years of professional activity.

We have now, however, to keep to our pressing problem of saving the memories of the Agnelotti back garden to Giovanni. We have to resist the temptation to leave aside the back garden to start working on these tickling intellectual challenges. The mathematical problems can wait; the back garden will be gone a few days from now.

We have to postpone our errands on the mathematical, metamathematical and metametamathematical characterization of the *CSP*. I am glad we wrote down in our scribbles what *CSP* stands for; I confess that I sometimes forget.

There is a point, however, that cannot be postponed.

Does it matter to consider that our preservation of the Agnelotti back garden is being prepared *for Giovanni*? If we were preserving the memory of the Agnelotti back garden for another person, would the result be different?

It certainly makes a difference to consider that we are only engaging in all this effort because Giovanni is Dr. Agnelotti's grandson. We consider that the information about the Agnelotti back garden will be considered relevant by Giovanni himself.

When he gets to the appropriate age, of course.

This age, in the case of Giovanni for the purposes of appreciating the back garden is precisely an age we have no idea which is.

Our objective rendition of the Agnelotti back garden, we suspect, depends on the reason it is being constructed, and on who is supposed to perceive it.

Wait a minute: are we *constructing* impressions about the back garden? Were we not supposed to *record* existing information about the back garden?

What are we talking about, the back garden or ourselves? Is Giovanni ever going to learn anything about the back garden? Are we not standing in between the back garden and Giovanni?

Could it be that we were actually employing this historical incident – the imminent destruction of the Agnelotti back garden – as an opportunity to talk and talk and talk to poor Giovanni about ourselves, showing off how smart we are?

Would Giovanni learn anything at all about the back garden from BackGarden version 1.0? Would he see two pedantic scholars turning their backs to the back garden and talking about themselves?

Of course not. We are doing our job, that's all. As best we can.

What Giovanni are we talking about, anyway? Who is this Giovanni, so ready to judge our work and ourselves? Is Giovanni-at-the-age-of-four the same person as Giovanni-at-the-age-of-six, or Giovanni-at-the-age-of-forty?

How many Giovannis exist? How many Giovannis must we take into account?

Does Giovanni-at-the-age-of-whatever need to know who prepared the BackGarden version 1.0 for him? Will it make any difference for him to know who prepared it? Would he consider differently the information contained in BackGarden version 1.0 if he were informed that it was prepared under the coordination of Dr. Agnelotti-at-the-age-of-sixty, instead of Dr. Agnelotti-at-the-age-of-twenty-five or Dr. Agnelotti-at-the-age-of-eighty?

How many Dr. Agnelottis exist? Who is preparing BackGarden version 1.0?

Who could imagine that a back garden could bring so many headaches to us?

Us?... Us who, after all? Who are we?

Are we, the proponents of this careful rendition of a back garden, the same two persons today as we were yesterday? Who are we? What are we talking about? What were we talking about yesterday? Who was here yesterday, in this office, working on BackGarden version 1.0?

Oops! Who was here yesterday sitting on this chair? Who was using *my* wallet and spending *my* money? Worst of all, who was here yesterday eating *my* tomatoes under the fig tree?

OK we are stuck. All we know is that the back garden exists; it has an existence of its own. We concede to that. Or at least take that by hypothesis. A working hypothesis. For the moment at least.

Assuming as a provisional hypothesis that the back garden does exist, we can also hypothesize that it bears genuine manifestations of its own.

OK fine. So the back garden exists and manifests itself in a multitude of ways. Or at least so we believe.

By "back garden", as we hope is perfectly clear by now, we mean a lot more than simply the present physical attributes that characterize its three-dimensional manifestation at this very moment.

The Agnelotti back garden comprises the fig tree. It comprises all symbolic meaning the fig tree brings to the Agnelotti. It comprises all memories and experiences related to this fig tree. It comprises all knowledge the Agnelotti have accumulated and experienced about fig trees, thanks to this specific fig tree.

The Agnelotti fig tree conveys memories, emotions and experiences of generations. The Agnelotti fig tree is Gigino, Cicilla, and everybody whose existence has become connected to this tree. Myself included. The Agnelotti back garden, to some extent, is me. I am, to some extent, the Agnelotti back garden.

The Agnelotti back garden shouts aloud uninterruptedly to all creatures that have ears attuned to the appropriate frequencies. It floods the world with streams of visions, facets and perspectives from which it can be seen. It shouts about itself. It shouts about us.

Here we are, Dr. Agnelotti and I. Two pairs of ears, listening to the Agnelotti back garden.

When we transcribe our impressions to BackGarden version 1.0, it seems that we need to add a warning to it: "Agnelotti back garden, as perceived in such-and-such day/month/year by Dr. Agnelotti himself, with the assistance of Dr. Agnelotti's colleague, etc. etc."

In other words, we are not really just simply and purely talking about the Agnelotti back garden. We are talking about ourselves. As seen from the back garden. As we see us being seen from the back garden.

When we describe the back garden we also talk about these specific two people who happen to be in this office in this specific moment of this specific day. Physically, these two blokes look quite like the other two who were here yesterday. They are not the same ones who were here yesterday, though, as seems so obvious now. They are coming here for the first time today.

Every morning two different persons come to work in this office. Strange how the security system at the Institute lets us in, despite this being our first visit here. Strange how they do not realize that these two strangers – us – are using the identification of those people who were here yesterday to sneak into the office of those two other people who were here the day before yesterday.

We are actually very different every moment. There is a different pair of "we" working on BackGarden version 1.0 each second. There is a different back garden each second. We must anchor and freeze our views somewhere, lest we will continue to wander adrift like this and not be able to do anything useful. The cost of anchoring and freezing is giving up some of the visions, facets and perspectives offered by the back garden.

OK, let us keep up with some loss then. It is going to be the Agnelotti back garden, from the Pleistocene until today, including all memories and facts and events that have occurred in the back garden since then. Or at least some of its selected manifestations since then. We shall not consider what happened before the Pleistocene, or what is going to happen from now on.

Let us freeze these two guys who are here today working on BackGarden version 1.0. Let us consider them from their birth until this moment.

Let us imagine all possible future Giovannis, and from them select one sequence of Giovannis. Mathematically speaking, let us consider a linear chain of Giovannis, such that we do not find two different Giovannis living at the same time. Different moments, different Giovannis, single moment, a single Giovanni.

From the selected chain of Giovannis, let us select that Giovanni who is blowing the candles on top of his five years old birthday celebration.

BackGarden version 1.0. From the Pleistocene until now. Includes all memories, events, emotions, as well as physical events. Includes the intentions, prejudices, points of view, opinions and interests of the authors.

BackGarden version 1.0. Prepared by Dr. Agnelotti, with the assistance of a colleague. Includes Dr. Agnelotti himself, as well as his colleague. As viewed by themselves. Their perceptions of themselves being influenced by their intentions, prejudices, points of view, opinions and interests.

BackGarden version 1.0. Specially prepared and tailor made for Giovanni on his five years old celebration. As imagined by Dr. Agnelotti and his colleague. In perfect accordance with their intentions, prejudices, points of view, opinions and interests.

We have done our best to withhold to the scientific objectivity of facts, in order to preserve the essence of the Agnelotti back garden with the utmost fidelity.

There are indications that the word *History* derives from the Sanskrit *Veda*, which can be translated roughly as *wisdom*, or perhaps *knowledge* depending on context. It was apparently the Greek scholar Thucydides, who lived in the fourth century *BC*, the first one to conceptualize History as the recording of facts based on a neutral point of view and controlled by the objectivity of purely neutral observations. From our personal experience, Thucydides has never undergone the experience of loosing his back garden because of some director plan. He would have learned, had he suffered that experience. He would have given a second thought to this "purely neutral observations" rubbish.

Dr. Agnelotti has the word. He has the right to it. After all, he is the grandfather. It was because of Giovanni that we started all this.

I try my best to capture the essence of the back garden, based on Dr. Agnelotti's descriptions.

Dr. Agnelotti starts by stating that the back garden is a very special place for him and his family. I write this down. I need to know what exactly "special" means in this context.

Dr. Agnelotti tells me the back garden is almost like a sacred ground for them.

I understand that "very special" is equivalent to "almost sacred". I wonder about the measurability of sacredness. If A is sacred and B is almost sacred, then it is fair to entail that there can be a C such that C is almost almost sacred, and that therefore A, B and C can be ordered according to their sacredness. Hence, sacredness determines a metric space.

For the time being, we avoid discussing the notion of sacredness. We consider the metric space determined by sacredness, whatever can be meant by "sacred".

I explain to Dr. Agnelotti that we may need to explain a little further the equivalence of "special" and "sacred", and that it worries me a little that we may need to theorize about sacredness.

Dr. Agnelotti laughs.

That was not meant to be taken literally. Yet, there is a mystical dimension to it. The ineffable is what makes that space so special. The back garden is made of sensations, emotions and memories, much more than trees and plants and walls.

Once the walls and trees and plants are gone, however, the back garden will be gone. So the back garden is also made of trees, plants and walls.

Could there be further substances that would be necessary to complete the description of the back garden?

Let us consider for a moment the fig tree. There are botanical features about it that can be mentioned, like its age, the environmental conditions to which it has been exposed, some morphological features – height, width, and so on – and its health conditions, to name a few.

We can extend our observations about the back garden this way, cataloguing each object that belongs to it, together with its most remarkable and evident features, including historical features that explain to some extent why each object is today the way it is.

The fig tree is also intimately related to other objects that have at some point belonged to the back garden, or at least been there. Like myself, for example. Like Gigino. Like Ms. Conchetta. Like the squirrels who occasionally have visited that tree.

The squirrels have always fascinated Cicilla. Indirectly and by transitivity, therefore, the fig tree has built a relationship with Cicilla, every time a squirrel has climbed it, and most remarkably in those moments when Cicilla could watch a squirrel on top of the fig tree.

We can thus also catalog the relationships among the different objects that at some point and to some extent have related to the back garden, thus helping to construct it along the years.

The fig tree also relates with history, and therefore with time. It tells stories and the history, for example, of the Agnelotti. There is that branch from which Gigino fell, right in front of Dr. Agnelotti, who rushed all the way to the infirmary with his son on his arms. There is that lower thick branch on which Cicilla used to sit to read.

OK, we can look for appropriate data structures to represent all that, and then build the general representation format, and finally fill in some forms, do some typing on the computer and *voila*! We will have captured the required information to rebuild the back garden. At least in virtual reality.

BackGarden version 1.0.

I smile proudly to Dr. Agnelotti. Dr. Agnelotti sighs and invites me to have a cup of coffee and make a break.

Something is missing. I cannot figure out what. My rendition of the back garden seems very close to useless, considering its goal.

Dr. Agnelotti tells me we need poetry. Without poetry we will not be able to capture the essence of the back garden.

I feel terribly impatient and exasperated. I can hardly believe in what I have just listened. How can poetry and scientific objectivity come together?

We need clarity, not poetry. Yet Dr. Agnelotti insists. Poetry. This is what is missing.

OK fine. Let us write down the history of pests that have affected the fig tree along the years, and then relate the pests with seasons and with the morphology and size of the trunk. Then we list the dates when Gigino fell down the fig tree and, through seasons and years and decades, build the relationships between the growth of the fig tree and Gigino's misfortunes, bruises and scars. Once this part of the work is done, then we look for some rhymes so that we can describe these relationships in verses, and it all becomes mystical, almost sacred and very special. As simple as that.

Maybe I do not understand what Dr. Agnelotti means by "poetry".

Could it be that form determines meaning in a message? Does the content of a text change when we write it as a poem?

Do the physical attributes of the back garden influence its significance for the Agnelotti?

Remove the fig tree from the back garden. You would remove with it a lot of experiences. You would remove part of the essence that defines the Agnelotti.

Form and content. Structure and meaning.

I thought I could study them separately.

I thought I could study them on another occasion.

Poetry.

This is what is missing.

The rough, obvious and unimportant aspects of the back garden do not require poetry to be captured. Giovanni-at-the-age-of-five, as we conceive him now, will find little need or interest in learning these aspects of the back garden.

We believe that the important aspects of the back garden for Giovanni-at-the-age-of-five are going to be the subtle ones. The ones related to sensations, feelings and emotions. These need a more refined language to be captured than the language that is used to describe physical phenomena.

They require poetry.

The plain language of sheer facts can characterize certain events and certain relationships. They can characterize for example physical attributes of objects and their relationships. They have a certain *granularity*.

Languages are like sieves. Every language can hold some events, but leaves other events – the subtler ones – pass through it. If we want to capture the subtler events, we need more refined languages.

The poetic dimension adds to the language of sheer facts the capability to hold subtler events, namely events related to sentiments, emotions and sensations.

We want to capture these subtler dimensions of the Agnelotti back garden. We nevertheless want to be disciplined, so that we make sure that no important aspect of the back garden is left behind.

We also want to be as objective as possible.

We need to find the way to capture in some sort of representation what we consider essential about the back garden. This must comprehend the relationships of the Agnelotti with the back garden. This of course includes the relationships of the Agnelotti with each other, and with other people.

We are going to capture in our representations what *we* consider is essential about the back garden. Hopefully, our perceptiveness will be good enough so that what is *really* essential is included in what *we* consider is essential.

Through BackGarden version 1.0, Giovanni is going to learn a little about the Agnelotti back garden. Hopefully, through BackGarden version 1.0, Giovanni is going to learn many things about the Agnelotti, which is what matters after all. Inevitably, through

BackGarden version 1.0, Giovanni is going to learn a lot about Dr. Agnelotti and myself.

With a little patience and attention, Giovanni is going to understand what we thought about him when we prepared BackGarden version 1.0, as well as what we thought he would be like at the age of five.

Giovanni may even have the chance to learn a few things about him, when he is confronted with the impressions he caused on people when he was six months old.

BackGarden version 1.0.

Time to head back from the Institute. Then dinner, then off to the ice cream shop.

Today we believe we have earned our right to the ice cream.

We have not written a single line describing the back garden yet. But we have learned many important things about the problem. We can be proud of ourselves.

Chapter 4

Shared Understanding

"Oh yes, yes, well said, that is how one does one's duty. Well, for someone who does not exist, you seem in fine form!"

Italo Calvino

One should not mistake shared understanding with shared knowledge. It is helpful, in order to clarify the difference, to consider the etymology of the words *knowledge* and *understanding*.

The origin of the word *knowledge* can be traced back to the Greek *gnosis* or to the Sanskrit *gnana*. In either case, it refers to direct and unmediated contact with reality. The world *understanding*, on the other hand, refers to *standing close* to reality (instead of *standing under* reality, as a hasty analysis of the word could suggest).

Shared knowledge, therefore – as will be further discussed in the chapters to follow – refers to mutual acknowledgment of integration with reality. Knowledge sharing deems information flow unnecessary, or at least redundant, since it only occurs when a group of agents is already integrated with reality – or at least with a common portion of reality – and, through reality, with each other. Shared knowledge presupposes mutual identification, which leads to a sense of intimacy. Information flow occurs when agents choose to stick to partial perceptions of reality for a purpose, to reach a goal or based on individual intentions. When a group of agents share mutual identification and intimacy, they also share their purposes, goals and intentions, as well as their views of the world, which do not need to be communicated among each other.

Shared understanding, on the other hand, refers to agents building equivalent artificial information systems as result of individual perceptions and information flow. Shared understanding, therefore, is a *result* of information flow. Shared understanding is a direct consequence of information flow. It cannot be built without it, and the reason agents may want to build a shared understanding about a certain portion of reality is to enable some efficient modality of information flow, namely the one that will ensure efficient satisfaction of the goals and intentions of the agents involved in the communication process.

Putting these two together, we find that knowledge sharing makes shared understanding unnecessary – or at least redundant. Our proposed view is that knowledge sharing is in itself the result of a process, instead of a tool to achieve some result through any specific process. Shared understanding, on the other hand, is precisely a tool that can be used in different situations and processes to obtain different results. Shared understanding is an instrumental concept, whereas knowledge sharing is not, in the sense that shared understanding is developed, pursued and refined for a specific purpose, and knowledge sharing is itself a purpose to be achieved.

Dialogues for shared understanding

Previously we have considered chains of agents acting as sender agents and receiving agents communicating information systems. We now turn our attention to chains that form a loop, i.e. any chain comprised by n agents $A_1 \ldots A_n$ and such that $A_1 = A_n$. To simplify our discussion, we consider the case when $n = 2$, namely the case when two agents engage in a *dialogue*.

Typically, in a dialogue agents take turns as sender and receiving agents. As before, a sender agent builds an artificial information system based on its perceptions, and communicates it to the receiving agent. The receiving agent in turn builds a second artificial information system, based on its own perceptions, that include the sender agent, the communicated system, a reflexive view of itself and any other perceptions of the reality the receiving agent may have – which are independent of the perceptions of the reality that the sender agent had at first place. Then the receiving agent assumes the role of a sender agent and communicates this second artificial information system to the first agent.

Each agent has its own perception of the reality, of itself and of the other agent. Each agent will have its own perception of each artificial information system being communicated. Hence, there is no reason to assume that a dialogue will necessarily converge to a single information system.

To illustrate this point, we can imagine a perverse thought experiment, in which a non-native speaker of a language – say, English – is taught purposefully mistaken meanings for each noun in English. Depending on how cunningly this shuffling of meanings is done, this speaker can engage into extensive dialogues with native speakers, in which each person is talking about completely different things, and yet assuming that the other is speaking about the same things as him.

The more common – and expected – situation, however, is that through dialogue agents build equivalent artificial information systems, in the precise mathematical sense that to each token in a system corresponds a token in the other system, and that to each relation in a system corresponds a relation in the other system.

Many social relationships – indeed, the vast majority of them – are built based on shared understanding. In other words, when we approach other people to engage into social relationships, we most often than not do so for a purpose, to reach a goal and based on our intentions. We then offer to those people not ourselves, not our perceptions of reality, not our perceptions of other people, but already filtered views of the reality, to ourselves, and to others, biased by our interests, plans and goals.

The people with whom we interact strike back with their own biased views, tailor made to ourselves and based on their interests, plans and goals. We are – or at least should be – aware of that, and so we take the artificial information systems we receive as such. We know that what we are receiving does not necessarily conform to reality; rather it reflects the distorted views of reality promoted by other people so that we can be used to help them reach their goals, based on their intentions and purposes.

This game goes through an initial stage when the proposed artificial information systems do not match with each other, causing some disruptions in the communication. These disruptions – commonly called *misunderstandings* in our everyday life – identify broken expectations of interpretation of pieces of information belonging to the artificial information system built by agent A, at some point in which it refers to an artificial information system built by agent B and communicated to A.

Successful social interaction occurs when the mismatching gradually decreases between the expected and the actual communicated artificial information systems. Yet, the resulting social interaction will be based on artificial information systems, which are built for specific purposes and to reach specific goals. When the goals are reached and the purposes change, the corresponding artificial information systems become obsolete and must be either discarded or revised.

As a consequence, relationships based on shared understanding – and therefore on artificial information systems – are by design ephemeral and remain alive only and precisely until their designed moment of expiration. Business relationships, social relationships and family relationships can be built entirely based on shared understanding, and this can be the explanation why we see disruptions in these sorts of relationships so often in our information based society and economy.

Shared understanding, nonetheless, is what occurs when equivalent artificial information systems are built, and the corresponding mappings among these artificial information systems are also built by each agent engaged in a dialogue. For all practical purposes, shared understanding enables a group of agents to perform in coordinated fashion, so that they can collaborate with each other to reach their individual goals. Shared understanding is a pre-requisite for cooperative work among many agents.

Communication spaces

We propose here a model to simplify the description of our proposed notion of shared understanding. Prior to presenting our model, we stress that it is *only* an artifice we employ to explain our view, and in our opinion should not be given further existence and importance than that.

Our artifice is founded on the assumption that there exists *a priori* an artificial information system, to which ideally the artificial information systems built by the agents engaged in a dialogue shall all converge. Initially, the agents have each its own artificial information system, built based on its perception of reality and assumptions about itself and the other agent. Through dialogue, the different artificial information systems are incrementally adjusted, in such way that a part in each artificial information system becomes incrementally more stable (i.e. resilient to change as result of the dialogue). These stable modules within the different information systems are the ones to become closer and closer to algebraically equivalent information systems.

All things going well, a dialogue shall, among other consequences, lead to the collaborative construction of modules within the artificial information systems being communicated between the agents participating in the dialogue, which shall be asymptotically equivalent. Our artifice amounts to the assumption that the artificial information system to be obtained as the limit case in this asymptotic process (in any of its equivalent forms) can be characterized *a priori*.

We call this information system that is built collaboratively by the agents engaged in a dialogue the *communication space* built by the agents for a specific dialogue. A communication space is an agreed upon configuration of terms to be shared by a group of agents, through which they interact in such way that the consequences of the interaction are (at least partially) predictable for all members of the group (Figure 4.1).

Consider, as a concrete example, the interaction between students and a lecturer along a teaching term at the university. During the first moments of the very first lecture, the lecturer and the students do not know each other. Presumably, the lecturer can make some assumptions about the students, considering the point in the course they are (e.g. second year students majoring in Computer Science); the students can also make some assumptions about the lecturer (e.g. based on his curriculum, or on the opinion of their colleagues), and both lecturer and students can make assumptions about the topics to be covered during the academic term. It is perfectly possible, however, that these assumptions do not match at all with each other, or with reality.

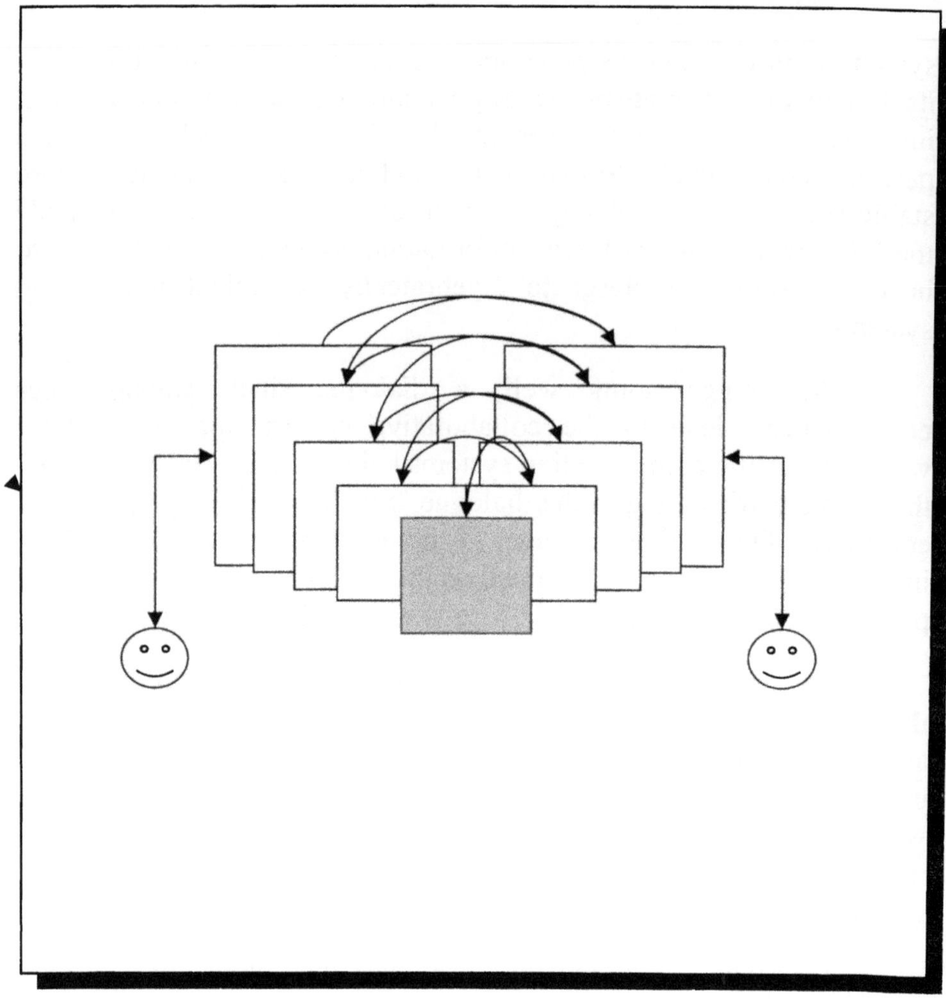

Figure 4.1. Collaborative construction of a communication space

As the academic term goes, the students and the lecturer interact during their lectures, through assignments, lecture notes, etc. All things going well, by the end of the academic term the lecturer and the students will have built partial views about each other and about the topics covered during the lectures, which will at least trigger appropriate behavior during interaction. For example, when the lecturer presents a problem to a student, the student shall make use of the appropriate topics covered during the lectures to solve it, and present the correct answer to the lecturer. Notice that it is not necessary that the student and the lecturer have the same *understanding* of the topics used to solve the problem, or of the problem being solved. As can be frequently observed in practice, it is also not necessary that students have the same opinions about a lecturer as the lecturer has about himself, nor that the lecturer has the same opinion about the students as they have about themselves. It is only required that these opinions trigger compatible behavior during the interactions between the lecturer and the students.

By the end of a successful academic term, the lecturer and the students must have built collectively a communication space. This communication space characterizes the portion of the technical information delivered by the lecturer that has been effectively absorbed by the students.

When all agents in a group access the same communication space (*modulo* equivalence among artificial information systems), they can engage in a *productive dialogue*.

Productive dialogue

A productive dialogue is such that the participating agents can coordinate their actions. The agents participating in a productive dialogue must not necessarily act together to reach a common goal. Indeed, they must not share any goal at all. What characterizes a productive dialogue is the fact that the participating agents in the dialogue have, thanks to the interaction with the other agents, more and better means at hand to reach their own goals.

There are two perspectives upon which productive dialogues can be analyzed:

1. The individualistic perspective, based on the perceptions of the dialogue realized by each individual agent participating in the dialogue.

2. The social perspective, based on the global result of the interactions among the agents participating in the dialogue.

Through the individualistic perspective, we analyze what can be achieved by an individual agent which would not be achievable for that agent without the participation of other agents. Through the social perspective, we consider the agents participating in the dialogue as a group, and analyze the result of their collective actions and perceptions.

What characterizes a productive dialogue is the fact that the participating agents intentionally acknowledge the participation of each other in the dialogue. This mutual acknowledgment is identified by the shared communication space built by these agents.

Productive dialogues are, therefore, the distinctive mark of social networks built for specific purposes and in order to reach specific goals. It is through productive dialogues that people interact in everyday life – in business or many of our routine social interactions – assuming the existence of artificial information systems similar to the ones they produce which, in the limit case, would comprise a communication space *modulo* equivalence relations, to function as the ideal mediator of goal driven social interactions.

Consider for example an electronic commerce scenario. To simplify our example, we consider the situation involving two persons, who take the roles of *buyer* and *seller* of a particular product.

Initially, buyer and seller do not know each other. The buyer has the specifications of what she intends to buy, and the seller has the concrete product he wants to sell. Based on the product itself, the seller broadcasts a description of the product, containing specifications, price, and any other features considered relevant by the seller considering his goal (to sell the product), his conceptualization of a prototypical buyer and his own perceptions about the product.

The buyer has the specification of the product to be bought, i.e. her perceptions about the product, her goal (to buy the product) and her conceptualization of a prototypical seller. She will look in the Internet for products and sellers who at least partially match her expectations.

Buyer and seller engage in a dialogue through the Internet to refine their views until they reach a coherent view about each other and the product they want to buy and sell. This dialogue can be synchronous or asynchronous, multi-pass or one-pass. It can be completely *ad hoc* or rely on off-the-shelf e-commerce templates and protocols through which the agents assume previously established roles to engage in the dialogue. The expected result of this dialogue is an approximation of a communication space that is acceptable to both participants in the dialogue.

After this approximation of a communication space is established, then buyer and seller can engage in a productive dialogue. Buyer and seller evidently do not share a common goal, but their individual goals can be reached e.g. if the commercial transaction is effected. The commercial transaction is a collection of coordinated actions through which these two agents reach their corresponding goals. In order for these actions to take place, each agent has to rely on actions of the other agent.

We can take the stance of any one of the individual agents, and analyze the commercial transaction as a process in which this agent interacts with an abstract environment – that happens to contain the other agent – whose behavior is to some extent predictable. This is the individualistic analysis of an electronic commercial transaction, through which we can, for example, build advice systems for individuals to behave effectively in order to reach their goals.

Alternatively, we can analyze the process as a whole, in order to understand for example the effects of electronic commerce in business organizations and macroeconomics. This is the social analysis of an electronic commercial transaction, through which we can, for example, identify good regulatory systems to detect and to avoid frauds, to improve the quality of life of all citizens in the long term, to reach social goals (as opposed to individual goals) such as insuring the preservation of cultural assets, or supporting the implementation of public health policies.

The social and the individualistic approaches to analyze dialogues and productive dialogues complement each other. When designing systems to enable dialogues, or analyzing *a posteriori* emergent systems and their consequences, these two approaches should preferably be both considered.

Facilitators for productive dialogues

More often than not, a dialogue is not purely genuine, as would be the case for example if a shipwreck occurred close to a distant island in the Pacific, a single survivor managed to swim and reach the island and met with a native citizen from that island. In the first moment, they could share nothing at all: language, cultural background, goals, interests, etc. and then after a lengthy process of interactions they could slowly progress to a rough communication space through which productive dialogue could occur.

The more usual situation is that the participants in a dialogue accept stereotyped roles for themselves, start the incremental construction of a communication space from commonly agreed upon off-the-shelf information systems, and also accept standardized and socially agreed upon procedures to interact in order to build a communication space with great efficiency.

This is precisely the case of the example in the previous section, related to electronic commerce. In this case, we have two individuals accepting to divest themselves from their individualities to become simply *buyer* and *seller* – two stereotyped roles known and deemed acceptable by both; start the construction of a communication space from commonly agreed upon information systems that describe to how a stereotyped buyer and a stereotyped seller see each other, their products of interest and the relevant potential interactions between them – namely, business transactions related to electronic commerce; and build a communication space very efficiently through the Internet based on standard communication protocols used in such situations and through digital communication technologies.

Each individual has to accept many constraints to behave according to their agreed upon roles, information systems and interaction protocols. Each individual has also to assume that the other participants in the dialogue are going to do the same. Finally, each individual must assume that the right roles, information systems and interaction protocols have been taken by all participating agents. There are many assumptions, therefore, to be made by each agent about it and about the other agents. These assumptions are not devoid of risk, and therefore any dialogue that is not purely genuine – in the sense of being built from scratch by the agents participating in a dialogue – is pervaded by uncertainty. We develop this topic further in Chapter 7.

A *facilitator* for productive dialogues is a conceptual tool made of three components:

1. A set of stereotyped roles for the participants in the dialogue.

2. A set of initial information systems for each of the participants in the dialogue, to bootstrap the construction of the communication space through which the productive dialogue shall occur.

3. Standardized procedures for interaction in order to incrementally build the communication space for the productive dialogue.

Facilitators for productive dialogues can be implemented in a variety of ways, depending on their intended use and users. One possible way to implement facilitators for productive dialogues is as software applications. We devote in this book some special attention to facilitators that can be implemented as software products.

A facilitator for productive dialogues is a *knowledge artifact*. An *artifact* is an instrument built by an identifiable author with a specific purpose in mind. Usually, artifacts are also objects that go through a well defined *design* process to be effectively constructed. Necessarily, artifacts are built using appropriate *materials*. A knowledge artifact is an artifact whose primary material is commonly named "knowledge". We accede to the proper name "knowledge artifact" in this paragraph only, to make the appropriate link to the existing literature; however we save the term "knowledge" for a specific notion, to be developed in the final chapters of this book.

Indeed, according to our terminology, the most precise name for the artifacts we have briefly characterized above is *information artifacts*. An information artifact is an artifact whose primary components are information systems and procedures to transform them. These procedures can be implemented for example as finite state machines, i.e. directed graphs whose nodes are sets of information systems and whose edges represent legal transitions between sets of information systems. In the case of facilitators for productive dialogues, we can for example use a finite state machine to implement the interaction protocols between agents to work collectively and converge to a communication space.

The point to be highlighted here is that information artifacts are first and foremost artifacts, and therefore proposed by an author for use by a group of agents to reach a certain goal which includes the communication space the proponent of the information space has in mind.

One important concept applicable to any artifact is the notion of *affordance*. By affordance here we mean the selected attributes of an object that influence how it is actually used. A classical example is the format of a handle on a door that determines how the door is opened – by pulling, or pushing, or twisting, and so on. In the case of information artifacts, we say that affordance is present when the users of an artifact can only use it the way that is expected by the author of the artifact.

In the case of facilitators for productive dialogues, a facilitator presents affordance if the interacting agents necessarily assume roles that are compatible with the roles proposed in the facilitator; the bootstrapping information systems are equivalent to information systems that are effectively adopted by the agents; and the interaction protocols used by the agents are (mathematically) equivalent to the protocols present in the facilitator.

Affordance is an important attribute to limit the uncertainty that is present in any interaction, including productive dialogues. It should be noticed, however, that it imposes operational constraints in the use of an artifact, not necessarily helping to ensure that the agents participating in a dialogue actually share a common perception of any piece of the reality. As an analogy, we can consider a hammer, which is an artifact whose affordance ensures that the user is going to hold it by the handle and hit the target with the head – hence that all uses of this artifact can be deemed operationally equivalent. The same hammer, however, can be used in very different contexts, e.g. to fix a nail while building a doghouse or to smash a walnut to be eaten. The proper use of a hammer by two different agents does not imply any shared perception of the reality, shared goal or shared understanding of their actions. All it implies is that operationally these two agents behave in very similar ways.

Bibliographical notes

The notion of knowledge outlined in the beginning of this chapter, founded on a brief etymological analysis, echoes the notion put forward by some philosophers throughout History. In particular, we find a notion very similar to this one in (Kant, 1784), however named as *Enlightenment* in that text.

The different viewpoints from which social interaction can be analyzed – which amounted to the individualistic and the social perspective of productive dialogues – are discussed in (Wenger, 1998).

A recent analysis of the role of artifacts in our everyday activities, focusing specifically on artifacts made of information, is developed in the PhD thesis of Colombo (2005). A more general analysis of artifacts and how they impact our lives can be found in (Norman, 2002). It was indeed this book that popularized the notion of *affordance* employed in this chapter.

Intermission

Today is the day. We have decided that today we start coding our representations of the Agnelotti back garden. We still do not know what is going to be part of these representations, or how exactly we are going to encode them.

We have decided to start coding it today, nonetheless, even if we prepare just rough sketches, otherwise we fear that we can continue to discuss about what and how and why to do things such and such way for a long time, and continue with no concrete work done. We do not have much time. We should better have something relatively good at the end of our work together, even if imperfect, than end up with nothing at all.

"After all, we really wish this work to be done by the time of Giovanni's fifth birthday party", as Dr. Agnelotti humorously stated during breakfast this morning.

Our starting point is free text stored in a computer file. We may include some multimedia material, like photos, drawings and sound recordings. The core representation is, nevertheless, text.

We have considered that this could be a fine shortcut to having concrete results very shortly. Considering all existing empirical evidence that written text can be very expressive indeed, as proven by universal literature, we are hardly going to be too far off the track with this choice.

Furthermore, we can work on representations of the back garden *in* the back garden.

Today we are not going to the Institute. Today, to my pleasure and excitement, we are working under the fig tree.

Immediately after breakfast, Dr. Agnelotti starts to tell me a story of when Gigino was nine years old.

Those were not very easy times, financially speaking. Dr. Agnelotti had to work twelve to eighteen hours a day, to pay the mortgage, the school for Gigino and Cicilla, and to help Mrs.

Agnelotti's family, who by that time had to pay for the hospital where Mrs. Agnelotti's mother stayed.

From Monday to Friday Dr. Agnelotti left home before six o'clock, to go to the Institute. He stayed there until four, when he left to the University to teach.

It was a small private University, and Dr. Agnelotti was hired as a free-lancer to teach some specific disciplines. Usually, Dr. Agnelotti had to take those disciplines no other lecturer would like. Dr. Agnelotti took as many disciplines to teach as he could. The University paid by number of lectures. Dr. Agnelotti usually stayed at the University until eleven in the evening.

The trains were not as good or as fast as today by that time. Dr. Agnelotti rarely arrived home before midnight. Then a shower, something to eat, and some sleep till five in the morning the next day.

Still the money did not break even by the end of the month. Dr. Agnelotti started to take some technical books to translate. He did the translation work on Saturdays.

On Sundays Dr. Agnelotti was dead tired, so he had to sleep almost the whole day. As a result, Gigino and Cicilla did not have much opportunity to interact with their father.

Dr. Agnelotti knew about the progress of his children thanks to Mrs. Agnelotti, who informed him about every event whenever she could. The children knew about their father also thanks to Mrs. Agnelotti, who explained to them that their father was working hard and all the time like that so that they could go to school and have some comfort. She explained to them that his work was beautiful and dignifying, so that they would respect their father as well as learn the importance of work.

One evening Mrs. Agnelotti surprised Dr. Agnelotti with the news that Gigino had asked her if he could work at the market once a week. The fishmonger needed an assistant and Gigino wanted to work for him.

This would not be an easy job. Mrs. Agnelotti said she also was surprised when Gigino asked for her permission to work at the market. She asked him if there was any particular reason for him to want that, and he explained that he needed some money because he needed to buy something. He could not tell what he was going to buy, because it was a surprise.

Dr. Agnelotti and his wife considered that Gigino was expressing with that attitude that he had understood something about the value of work. They decided to give him permission to work at the market.

The following week Gigino was at the market. He worked there every Wednesday afternoon. It was indeed a tough job, and Gigino arrived very tired home in the evening. Yet, he did not give up.

Gigino worked at the market for three months. After three months, he asked for permission to leave his job. He explained that he already had earned the money he needed to buy the object he needed.

Dr. Agnelotti thought that he would find a new football in his garden the next morning. Or perhaps some other toy.

Dr. Agnelotti was not prepared for the surprise.

On Friday morning, two days after Gigino had stopped working at the market, Dr. Agnelotti was leaving home to work as usual. It was half past five, and it was still dark. Winter was approaching. It was also a bit chilly.

Dr. Agnelotti almost bumped his head on that parcel that was hanging from the lower branch of the fig tree.

It was a soft parcel, anyway, but it definitely was not small.

Dr. Agnelotti saw that his name was written on it. He recognized Gigino's calligraphy. His name was written on the parcel, and nothing else.

Dr. Agnelotti was intrigued. He opened the parcel and found a beautiful cardigan. It was a very good cardigan, and probably a very expensive one. It was his size.

That evening, Mrs. Agnelotti told him that Gigino had woken up very happy. She had asked him whether he had bought that whatever he wanted to buy. "Not wanted, mom. *Needed* to buy", he corrected her.

She asked him what he had bought, and he told her that she would know by the time his father arrived home in the evening. He told her that his father worked too much and had to walk to and from the train station every day, and that he had to be warm during this walk.

When Dr. Agnelotti arrived that evening wearing his new cardigan, they both understood the whole thing.

Fortunately for the Agnelotti family, the Institute changed his status within the National Research Council six months after this story had happened. The salary of Dr. Agnelotti was substantially improved. A few months later, Dr. Agnelotti was offered a promotion, and almost immediately after that he got a permanent position at the National University.

Their time of financial penance had been left behind. Dr. Agnelotti continued to wake up and leave home very early in the morning, but then he could leave the National University at seven, and be back home by about eight in the evening every day. He did not have to do the translations on Saturdays, so the whole family could stay together, go to the beach sometimes, and so on.

Those were not so bad times, though. Mrs. Agnelotti had managed to transform the enforced absence of Dr. Agnelotti into a lesson to be learned by the children. Thanks to Mrs. Agnelotti magic powers, those times helped to strengthen the links among the members of the Agnelotti family. They also helped to build up the character of the children, and to make them all admire and love Dr. Agnelotti even more.

The fig tree and the back garden were not leading characters in this story. Yet, they were there. The fig tree had a fundamental role in the plot.

Indeed, the story started to be recalled today because we are here, under the fig tree.

The fig tree is the key that opened the gates of memory to bring this beautiful story to the present.

I now believe I am starting to understand the importance of the back garden to the memory of the Agnelotti family.

Computationally speaking, the back garden is a structured collection of pointers, each pointer associated to a sensation, a story, an aphorism, and so on. The objects which are identified by the pointers which are present in the back garden characterize important facets of the history and character of the Agnelotti family.

I explain this model enthusiastically to Dr. Agnelotti, immediately after he tells me the story of Gigino and the cardigan.

Now we are ready to start.

Dr. Agnelotti looks at Mrs. Agnelotti. They both sigh. Dr. Agnelotti asks if I would like another cup of coffee.

OK, maybe I have simplified things a little bit.

Mrs. Agnelotti leaves to the kitchen, to bring some more toasts. We are savoring toasts with the delicious plum jam they have prepared this year. The plums, of course, come from the garden.

They started planting plums about eight years ago. They did not know anything about plums. Essentially, they ate some plums they had bought at the market and then buried the stones. To their surprise, it worked.

A little later Mrs. Agnelotti bought a beautiful book about the cultivation of plums. They prepared the ground, replanted the small plants that were already there and planted a few more.

The trees grew beautifully. The first time they bore fruits the whole family celebrated. Year after year, the plum trees bore more and more fruits.

It became a tradition in the family to harvest plums for the jam. Every year the whole family waited the Sunday when they were going to catch the plums. It was always on a Sunday morning. The whole family would work together.

Catch the plums, and then wash the fruits one by one. Cicilla was always in charge of peeling the fruits. She did that very meticulously and carefully.

Slice the plums, and then put them to boil with sugar, lemon and some herbs. Mrs. Agnelotti would be in charge of the cooker. She had the patience to stir the jam continuously until the jam got ready.

Dr. Agnelotti and Gigino prepared the jars. The jars had to be sterilized to hold the jam. With properly sterilized jars, the jam lasted for as long as three years.

The whole family worked together to prepare the plum jam. It was a celebration. It became a ritual. One of the many small rituals the Agnelotti family held every day.

Even after Cicilla got married and Gigino moved to study in a different city, they still come every year to catch the plums and prepare the jam.

There is something about the plums and the jam that defines the Agnelotti family. There is something about the Agnelotti family that bears meaning to each jar of jam.

The back garden can point to the plums. The plums can point to the jars of jam. The jars can point to the yearly ritual of preparing the jam. The preparation of the plum jam can point to the subtleties of the relationships, sensations and feelings of the Agnelotti.

I can see the back garden. All I have to do is look around. I can build representations of it.

I thought I could see the plums. I can see some of them, over there, hanging from the tree. They are beautiful, by the way. However, these are not *the* plums. They are only a few plums. They are fine representatives of a family of plums. The plums that grew from the Agnelotti back garden. Present and past plums. These plums, on the other hand, are not worth for what they look like, or for what they taste like, or for the quality of the jam that can be made with them.

Which, by the way, is fantastic.

The plums are there to evoke sensations related to the family life and history of the Agnelotti. The plums are pointers to sensations.

Plums are also plums, wherever they are. Without physical plums, there would be no jam. In a very concrete sense, the plums point to the jam.

Savor a bit of jam. It is delicious; there is no doubt about it. Can we even imagine the taste of the jam as perceived by Dr. Agnelotti? I suspect we cannot. For Dr. Agnelotti, the jam tastes of family gathering, of laughs and songs and playful Sundays.

The jar I have in front of me points to many jars of jam, present and past. Each of these jars points to *the* jam, which points to something you can taste, as well as so many other sensations, feelings and history.

The jam points to the family reunions on those Sundays devoted to catch plums and prepare jam. Those family reunions point to family reunions in general. The Agnelotti family reunions point to the fundamental character of this lovely family.

Pick a plum from the back garden. Any plum. Look at it carefully. You should see the whole history of the Agnelotti in that plum. You should understand the character of the Agnelotti through that plum. You should be able to know the Agnelotti. You should be able to start loving the Agnelotti by looking at that single plum.

Giovanni will have to learn about the Agnelotti from plums he will not have the chance to see.

The whole story about the plums and the trees and the jam ritual must be preserved. The subtle meanings carried by this story must be preserved.

We all stay silent for a moment. Five minutes, if that much. It seems as if we stay silent for hours. I do not feel like moving, lest I would break the spell of that moment.

A moment of celebration in itself. A sacred moment. The magic of the plum jam unveils itself, we are involved by it.

Mrs. Agnelotti fills my cup of coffee. I savor it. It tastes different. For a moment I feel as if I had been there on Sundays, singing and making jokes and picking plums.

As if I knew a good plum from a sour one.

OK, so maybe the data structures need to be a little more complex. We need pointers and meta pointers. We may need higher order logical inferences to formalize the meaning of plums. It may take a while, but I believe I know how to sketch the formal model to capture the meaning of plums and jams. All we have to learn is how to represent the terminal nodes in this hierarchical tree of concepts.

There may be even a theory we can develop, relating the depth of a node in a tree of concepts with its degree of abstraction. First we may need to specify a measure of abstraction, somehow correlated to the difficulty to express a concept linguistically. The top nodes in the tree, for example, may be the less abstract ones – plums and trees and the back garden as it can be perceived physically around us. A botanist can describe these nodes with perfect clarity and precision.

The bottom nodes refer to sensations, feelings and those characterizing invisible features of the Agnelotti family. They can be perceived indirectly through stories and memories, yet we must find a more direct and unambiguous way to capture these notions in order to represent them.

Good, I think we have a model to start working. A pencil and a piece of paper would do well to start sketching something.

Again, Dr. Agnelotti and Mrs. Agnelotti sigh.

Mrs. Agnelotti tells me I have already had too much coffee. Maybe I would like some herb tea?

I sometimes have the impression that Mrs. Agnelotti offers me something to drink whenever she thinks that what I am saying sounds too stupid.

A funny thought, of course, and certainly an improper one, considering the kindness of Mrs. Agnelotti.

And considering that I am after all simply trying to paraphrase their stories in forms that could be more palatable to formal representations and computational data structures.

Assuming by hypothesis that BackGarden version 1.0 is going to be implemented as a computer system. We had not mentioned that explicitly before, but it seems the obvious choice.

It is nearly half past ten now. We are still under the fig tree. I feel a little distressed, we have not sat in front of a computer so far, and I feel as if we have no work done. I feel worried that we may spend the whole day just talking and having a good time away from the Institute. The more we talk the more I feel that our project is quite complicated. I wish we finished this long breakfast soon and start working.

On the other hand, I wish we could stay here all day. It has been so pleasant, this conversation. Strictly speaking, Dr. Agnelotti and his wife did not tell me so many concrete facts and events that could be added to data structures and symbolic representations to enrich BackGarden version 1.0. Yet, I feel as if I knew more now than ever about the Agnelotti, this house and this back garden. Oddly, I feel as if I have also become part of the stories I have been listening.

I feel a little worried and uncomfortable for having such a good time this morning. I feel as if I should be working, instead of drinking coffee and savoring toasts and jam.

Dr. Agnelotti does not seem to share my worries. Actually, he seems quite happy.

I suggest we take our laptop so that I can type some ideas. Mrs. Agnelotti sighs. I prepare my cup. There comes a little more coffee.

Dr. Agnelotti agrees with the laptop idea. I feel a little less stupid.

Mrs. Agnelotti asks if we could go first to the grocery and buy some lettuce for lunch.

For some unsuspected reason, I feel very happy with this alternative to getting the laptop.

Off we go to the grocery.

There is a little trick to close the front gate. You have to swing it in a specific way; otherwise it will not lock properly.

I swing the gate and it gets locked. I do it effortlessly. It does not seem such a noteworthy skill, knowing how to lock the front gate of the Agnelotti's house. For some unsuspected reason, I feel very proud I can do it.

As we walk to the grocery, Dr. Agnelotti laughs joyfully and starts talking about the front gate.

There was another front gate before this one. A beautiful gate made of carved wood. Quite heavy and extremely difficult to preserve, but it was so beautiful that Dr. Agnelotti considered more than worth the effort of varnishing it every year.

The original gate was already there when the Agnelotti moved to this house. No one seemed to know from where it came, or who had carved the ornaments it had in both sides.

Dr. Agnelotti breaks into laughter and tells that the original gate did not survive Cicilla learning to drive.

Cicilla had already got her driver's license. She had not had the opportunity to practice, though, and she felt a little uneasy every time she took Dr. Agnelotti's car. As Dr. Agnelotti explained, Cicilla did not consider herself a good driver at all, even though she could drive very well.

From what I know about Cicilla, I can imagine this was very true. Cicilla is always meticulous and careful about anything she does. She is the type of person that is never happy with less than perfection.

San Isidro never had much traffic anyway, so Cicilla could practice driving along the town.

It was a winter morning, and there had been some frost the night before. The floor was quite slippery, and it was still dark.

Cicilla decided to make a surprise to the family. She succeeded well in surprising everybody, including herself.

The plan was to take the car, sneak out quietly, drive to the market and buy some stuff for breakfast. She would be back before anyone would have wakened up, with a lot of cheese, bread and fruits. Of course she could walk to the market. It is less than ten minutes walking from the Agnelotti's house. Less than five minutes walking together with Dr. Agnelotti. But Cicilla wanted to practice driving.

What could go wrong with a plan like this?

Ms. Conchetta's black cat.

Cicilla got into the car. She had to drive backwards until she got closer to the front gate. Then she had to stop the car, go out, open the gate, take the car out of the garden, close the gate and leave to the market.

Cicilla turned the car on. And then it happened.

Ms. Conchetta's cat was sleeping under the car. With the noise, the cat got scared and jumped to the front of the car. Cicilla got scared by that ghostly black apparition jumping in front of her and momentarily lost control of the car. When she realized what was going on, she tried to use the brakes, but it was too late.

The car knocked the front gate down. The noise frightened Cicilla again. That time, she became definitely convinced that something spooky was going on outside the car. She tried to lock the doors and brake the car at the same time. She used the wrong pedal, and stepped furiously on the accelerator. The car screeched and speeded – backwards – downhill to the road in front of the Agnelotti's house. Cicilla screamed frantically inside the car. The cat accommodated itself on the corner to continue sleeping.

The whole family jumped to the bed and out of the house to see what was going on. They could see in the dark that the car was not there. Then they realized that the front gate was not there either. And then they realized that Cicilla was not together with them.

And then they heard the horn of the car. It sounded a little distant.

Then the car door opening, then shutting.

Then Cicilla's voice coming from what seemed as two blocks away. Delicate as always.

"Oops! Sorry!"

And then Dr. Agnelotti's steps running out of the house.

And then Dr. Agnelotti bursting out laughing.

The car was not much damaged, and it was fixed after three days. The front gate was ruined. This was how and when it was replaced by the one that is in use today.

I do not think I have met many people in the world who would react in a situation like this with a burst of laughter. Dr. Agnelotti did. With that, Cicilla felt his support, and did not feel so bad about the whole mess she had just created.

With small attitudes like that, Dr. Agnelotti and Mrs. Agnelotti slowly contributed to the definition of the personalities of Cicilla and Gigino. With small attitudes like that, they strengthened their links and expressed their love for each other.

Inexperienced drivers destroy front gates and walls and lamp posts to the hundred every day around the world. Cicilla ravaged the Agnelotti's front gate once, and it was many years ago. And that was a unique experience for the Agnelotti family.

Every moment has significance and importance. Every moment counts. Every inch has significance and importance. Every inch counts.

We reach the grocery and choose the lettuce to bring back home. I get distracted by the objects in the grocery and do not conclude my thoughts about this story.

I have the intuition that there is something very important in this story for our project. Something related to the old front gate.

I cannot find what it could be.

On the way back I start again. Maybe we can add a temporal dimension to our representation. We would then need a sorted system of pointers, of course. The algebraic structure of the semantics of this system may be a little complicated, but I believe we can organize it in such way that a decidable fragment of the syntactical counterpart of the language can be built. With a little luck, we will be able to express everything we want about the back garden within this decidable fragment of the language we will necessarily have to build.

Dr. Agnelotti walks silently. From time to time, he seems to hold his laughter. I suspect he is still living back his time with young Cicilla and the ruined front gate. I continue with the elaboration of my mathematical model, nevertheless. We have work to do.

We are approaching lunch time. Since today we are working at home, we all go to the kitchen to help in the preparation of the meal.

I look at the kitchen and it looks different now. Somehow the conversations and stories this morning changed the appearances of everything. I see things and objects and details I had not noticed before. I appreciate the things and objects and details I had already noticed with a different interest now.

What stories would the cooker have to tell me? What stories would the Agnelotti have to tell me about the cooker? How does the cooker relate with the plums, the fig tree, and the tomatoes?

During lunch I comment that I feel that everything Dr. Agnelotti and Mrs. Agnelotti tell me is important to describe the back garden. I also mention that I am finding it very difficult to structure this information in a very organized mathematical model. I wonder whether the structure of the representation of the back garden itself should reflect this fragmentation of the memories and sensations and feelings it evokes.

Perhaps the representation itself should evoke the same sensations and memories and feelings, instead of simply providing the means for the analytical understanding of what had occurred in the back garden along all those years.

Perhaps a better foundation for BackGarden version 1.0 would be a mosaic. Just like those beautiful stained glass windows we find in the cathedrals. Each story and memory and sensation is a piece of the mosaic. The conjunction of all pieces forms an image. The image is the desired representation of the back garden.

Dr. Agnelotti smiles. Mrs. Agnelotti does not offer me another glass of water.

Bingo!

Simple to state, difficult to execute.

We have at least now an agreed upon starting point.

This morning we started with completely different views about what the back garden was. It seems we also started with completely different views about what our work was.

I start to understand now. The stories and memories and sensations. They were part of the work. Pieces of the mosaic.

This morning I wanted to build a formal language, or perhaps a data structure, or perhaps an algebraic construct. I believed that this could capture the guts and innards of the back garden.

Dr. Agnelotti, however, wanted to build poetry. He wanted to write songs and wander around the world like a medieval minstrel.

I could not accept this approach to represent the necessary information about the back garden.

Dr. Agnelotti could not accept my rationalist approach to represent the essential information about the back garden.

It took me a few of the fascinating stories about the Agnelotti family to start changing my perception about the essence of the back garden we so much wanted to represent. It took me a few stories to start changing my understanding about the role of the representation and how the representation of an entity and the entity itself must be closely and intimately related. An entity and any of its representations are inseparable.

It took Dr. Agnelotti the observation of my appreciation of his stories, and the appreciation of my effort to transform those stories into concrete representations of the back garden, to accept that representations would have to be a little more concrete and formal than he was initially conceiving. Even if these representations were built to represent the ineffable.

We are not writing songs. I cannot play the lute, we are not minstrels, and we are not poets. Dr. Agnelotti has accepted that. I understand that this could indeed be the best way to preserve the back garden. It worked for the cultural heritage of the thirteenth century; it should work for the back garden. Giovanni would learn important things about the back garden and what it represented this way. Unfortunately, we are not poets, we do not play the lute, we are not minstrels.

We are not writing algebraic structures. The back garden defies mathematical rationalism. The subtleties and the ineffability of the emotions, sensations and memories of the Agnelotti defy logical representations. I have finally understood and accepted that.

We are not writing *about* the back garden. We are preserving the back garden itself.

Even though the back garden is about to be dismantled and turned into a road.

What is the back garden?

A mosaic. This is what we are building. A mosaic of small pieces. Each piece itself a mosaic. Memories, sensations, feelings. The Agnelotti is the back garden. The back garden is the Agnelotti.

We are building a mosaic. We add a small piece here, another small piece over there, and we see an image being formed.

I believe we are building a mosaic. I am not completely certain whether the mosaic is building itself.

Or perhaps Dr. Agnelotti is building the mosaic, and I am just helping. Dr. Agnelotti is building the mosaic, as a gesture of gratitude for the mosaic having built Dr. Agnelotti.

I put the pieces.

I build the mosaic.

I am a piece.

I am part of the mosaic.

We are pieces of the mosaic.

We build the mosaic.

As a gesture of gratitude for having been built by the mosaic.

The representation is the back garden.

The representation is us.

The representation builds us.

We build the mosaic.

We build ourselves on the way.

Chapter 5

Collaborative Problem Solving

"I keep to the rules. Do that yourself and you won't make a mistake."

Italo Calvino

Collaborative problem solving is a productive dialogue in which the participating agents have a shared goal and act collaboratively to reach this goal.

We are, therefore, now adding to our framework the notions of shared goal and collaboration.

A shared goal by a group of agents is actually not a single goal, but a collection of individual goals that are *synchronized*, i.e. when one goal is reached necessarily the others are also reached.

Collaboration occurs when an agent performs an action based on the expectation of another action to be undertaken by another agent. Collaborative problem solving requires, therefore, that agents rely on each other.

Collaborative problem solving accepts the presupposition that an agent communicates with others to reach a goal. Communication, therefore, is a prelude to goal driven action. The purpose of communication for collaborative problem solving is to *inform* action. Informed goal driven action is guided by the results of communication, and is therefore to some extent controlled by communication and predictable through communication. It is through communication that agents and their actions can have expected results and become, therefore, reliable. The actions of an agent can be monitored by the messages it sends, instead of directly by the perception of the effects of these actions.

Consider for example a medical emergency situation in a far off location, in which a nurse must perform a surgical procedure in order to save the life of a person who has suffered some sort of accident, counting on the expertise of a medical doctor who interacts with the nurse over the telephone. The nurse can describe what he is observing from the patient, based on which the doctor can advise the steps to follow in the procedure.

The doctor's actions are *speech acts*, i.e. illocutionary actions whose effects amount directly to the construction of a communicated artificial information system at the nurse's side of the telephone line.

The nurse's surgical actions are physical actions performed over the patient, i.e. actions that effectively transform the physical reality most directly experienced by the nurse and the patient. These transformations are perceived by the nurse through an appropriate information system, which is used as the basis for the nurse to build an artificial information system. This artificial information system is the object of other speech acts, which guide the construction of an artificial information system, now at the doctor's side of the telephone line.

Our electronic commerce example of the previous chapter can also be used again here. The individuals who have engaged in interaction based on the roles of *buyer* and *seller* build a communication space through which they have a productive dialogue. There is no single goal shared by buyer and seller: the buyer needs to satisfy some necessity through the product being acquired, and in order to do so he needs to perform certain actions, based on his purposes and a specific goal; the seller performs different actions, based on her own purposes and in order to reach a different goal. However, the individual goals of the buyer and the seller are synchronized (notice that the stereotyped roles of buyer and seller are simplified so that each of them has a single goal – buying a specific product for the buyer and selling a specific product for the seller) – one goal is reached if and only if the other is also reached. The agents must nevertheless trust each other, and trust that they have effectively managed to engage into a *productive* dialogue, to act – lest they become paralyzed by not being able to rely on the corresponding actions of each other, and fall into a deadlock as occurs in the Coordinated Attack Problem well known to computer theorists and logicians.

In very simple and anecdotal terms, the Coordinated Attack Problem goes as follows: two generals who cannot see each other have their troops camped on strategic sites close to a village controlled by the enemy. In order to conquer the village, the two generals must attack simultaneously, and hence they must coordinate their actions with perfection. General One will not attack until he is informed that General Two is ready for action, and vice-versa. The two generals know about each others' condition for action, and so each of them will only attack when he knows that the other general knows that he is ready. General One then sends a soldier to tell General Two that he is ready. When General Two gets this message, he sends a message back to inform that he has gotten the first message. General One of course will not attack until he gets this acknowledgment of receipt of the first message, and General Two knows that too. So General One must send to General Two an acknowledgment of the acknowledgement, otherwise General Two will not attack. Then General One will have to wait for the acknowledgment of the acknowledgment of the acknowledgment, and so on and so forth, and as an end result of this process General One and General Two are paralyzed for all Eternity, thus characterizing a deadlock.

To break the deadlock in the Coordinated Attack Problem, the communication protocol must be somehow bypassed. For example, one of the generals can bet that a message has gone through, and go for the attack trusting that the other general will do the same. In order to do that, the general needs *trust* and *self-trust*: he must trust on the other general, as well as on his own capability for action and for judgment of a situation. Two possible outcomes can result from the bet:

1. The second general bets that the first general will behave the way he did and behaves likewise, in which case the military operation is a success; or

2. The second general does not act synchronously with the first general. In this second case, collaboration breaks down and the operation fails. The expectation of occurrence of certain events in the reality, built through information systems, is not fulfilled – in other words, some information system proves to be imperfect and to not correspond with relevant events in the reality – and a culprit mismatch between the information system and the reality leads to improper behavior of an agent (the first general in our example).

A special case of interest in collaborative problem solving occurs when the agents participating in a productive dialogue share a single goal, i.e. when their shared goal is a singleton set. This is indeed the most common notion of collaborative work that is used in everyday language, which we are here generalizing. As an example of this special case, we can consider the simple situation in which five persons work together to carry a grand piano upstairs. They have to communicate, and their communication must be a productive dialogue, in order to coordinate their actions. They all must share a single goal – take the piano safely to the desired floor – which nevertheless must not be the single goal of each individual agent – only the persons at the lower stairs may have the additional goal of not being crushed by the piano in case it slips downstairs, for example.

Coordinated action

Coordinated action occurs when several agents act together, in such way that an action from an agent is based on acknowledgement of the actions of the other agents. Coordinated action is based on mutual expectations of actions.

These mutual expectations result from communication. More specifically, they result from *productive dialogues*, in which a communication space is approximately identified by the agents participating in the dialogue. The precision of the coordination is determined by how accurate the participating agents have identified their communication space.

Other constraints also determine the precision of the coordinated action involving multiple agents. Among them, we highlight the noise and distortion occurring in the communication among agents.

Collaborative problem solving requires coordinated action. Coordinated action, however, can occur without collaborative problem solving – if a group of agents does not have a shared goal, for example, they can still coordinate their actions and make use of the behavior of each other to reach their individual goals. This is frequently the case, for example, in corporate environments, where an individual identifies a certain psychological pattern and repertoire of standard strategies and practices in one of his colleagues, and decides to make use of it for his own purposes. These individuals are exchanging artificial information systems – even if it is not a purposeful act for one of them – and therefore are engaged in a dialogue in the sense we have proposed here. They are not, however, converging to a communication space, and therefore they cannot engage into a productive dialogue, which is a pre-requisite for collaborative problem solving. Some goals of the individual who is making use of the behavior of his colleague may be facilitated by the actions – and indirectly by the goals – of his colleague. There are not, however, goals belonging to these two individuals that could form a set that could be identified as a shared goal.

In other words, coordinated action can be the foundation of a variety of sorts of interaction. Collaborative problem solving – which we analyze in this book – is one of them, and the sort of interaction depicted in this example – which could for example be named parasitism, and shall be left for discussion in future works – is another one.

Coordinated action requires, nevertheless, *reliability*: each agent must behave based on expectations of the behavior of the other agents, and in order for the coordination to be successful these expectations must be fulfilled.

We briefly review here the concepts and terminology discussed thus far, pinpointing the potential points of mismatch between reality and expectations, which must be reliable for coordinated action to be successful. Initially we have simply *reality*, which includes the agents interested in coordinating their actions, their actions *per se* and the consequences of these actions, which are part of the natural dynamics of the reality.

In order to simplify our depiction of the relevant concepts, we consider from the reality the interaction between two agents, named *A* and *B*. The actions that must be coordinated occur in the reality. However, the coordination is effected through the mediation of information systems. Instead of the whole of the reality, including agent *B*, agent *A* selects an information system as the basis to interact with reality. The selected information system is an incomplete and possibly imperfect perception of the reality, and the more incomplete and imperfect the selected information system is, the less reliable the whole process becomes.

Then agent *A* builds an artificial information system to communicate to agent *B*, which is unlikely to be perfectly faithful to the selected information system. The more distortions are added to the artificial information system, which distinguish it from the original information system, the less reliable the process becomes.

Then the artificial information system is communicated from agent *A* to agent *B*, and the communication process adds further distortions to the artificial information system, e.g. due to noise and to limitations of the communication channel. The more communication distortion, the less reliable is the process.

Agent *B* takes into account the "artificial reality" proposed by agent *A* as the basis for its perceptions and decisions. Instead of experiencing the same reality that was initially perceived by agent *A*, agent *B* takes into account the received artificial information system, itself and the sending agent, i.e. agent *A*. Based on this reality, agent *B* selects an information system to employ in its decisions, which is an incomplete and possibly imperfect rendition of this "artificial reality". This new source of incompleteness and imperfection is another factor to potentially decrease the reliability of the process.

Agent *A* performs an action. This action is triggered not exactly by reality, but by agent *A*'s perception of reality characterized in the information system it selected. Agent *A* also informs agent *B* about this action, through the chain of artificial information systems referred to above. Agent *B* wishes to coordinate its actions with the actions of agent *A*, but instead of this it coordinates its actions with its perception of the actions of agent *A*, as communicated by *A*. The more imperfection and incompleteness is added all the way through, the less reliable the process becomes.

Consider flirting. Let us now make the example above a little more personalized and anecdotal. Instead of agent *A* and agent *B*, let us consider *Andy* and *Betty*. *Andy* wishes to cause a good impression on *Betty*, because he is romantically interested in her.

Actually, *Andy* is not quite interested in *Betty*, but in some information he has perceived about her that has made him believe that she can fulfill some of his needs – which can be physiological, status related, emotional, or social. *Andy* then builds an artificial information system about himself to communicate to *Betty*. *Betty*, on her turn, perceives the communicated information system, together with her own impressions about *Andy*, and builds her own conclusion – which, by using our terminology, is an artificial information system she builds to communicate to herself something about the reality she has perceived, already contaminated by her interests, prejudices, purposes and goals.

It is then *Betty*'s turn to build an artificial information system tailor made for *Andy*, which is communicated to him. Based on the signals *Betty* and *Andy* send to each other through their custom built artificial information system, they may at some point conclude it is time for action. This is the point when one – or both – of them makes a bolder move to hands being held, a hug, a kiss, and so on.

Clearly, *Andy* and *Betty* do not act based on the reality, but on exchanged artificial information systems that convey impressions of themselves fully pervaded by intentions and interests of each other. Miraculously, quite often these partial views of each other suffice for a couple to engage into coordinated action that leads to a romantic relationship, and sometimes even marriage and the beginning of a family.

Families and relationships based on information exchange and coordinated action only, however, have proven to be rather fragile. Coordinated action is founded on the satisfaction of individual needs that are not necessarily shared, and on the notion of mutual reliability in order to consider the coordination of actions a productive attitude. Once these needs are satisfied or considered no longer interesting by the agents engaged in this sort of relationship, or an agent behaves unexpectedly, thus exposing the unreliability of the relationship, the whole relationship becomes uninteresting, and then the participating agents may align their behavior to the tenets of rationality, and simply decide to break the relationship.

We have been careful not to confuse the notions of *reliability* and *trust*. All concepts discussed thus far to enable coordinated action are based on reliability. Reliability is a neither a necessary nor a sufficient requisite for trust, likewise for trust with respect to reliability.

Reliability and trust

Our proposed notion of reliability is based on the notion of *predictability*. Consider agents *A* and *B* which are engaged in some sort of interaction aiming at the coordination of their actions. Consider, for the purpose of presentation of our ideas, that agent *B* is the one to make efforts to coordinate its actions with those of *A*.

The possible history of one step of interaction is depicted in Figure 5.1. The numbers linked to the arrows indicate the sequence of events, which goes as follows:

(1) Agent *A* collects and structures its perceptions of an information system proposed by the portion of the reality that is available to it.

(2) Based on these perceptions, on its perceptions about agent *B*, based on an artificial information system published by *B* to *A*, and on its perceptions about itself, which includes a model of its goals, purposes, plans and beliefs, agent *A* builds an artificial information system to be communicated to *B*.

(3) Based on its own perceptions about the portion of the reality that is available to it, on its perceptions about itself and on its perceptions about *A*, which are based on the artificial information system communicated in (2), *B* perceives an information system, which is the one containing information about the expected behavior of *A*.

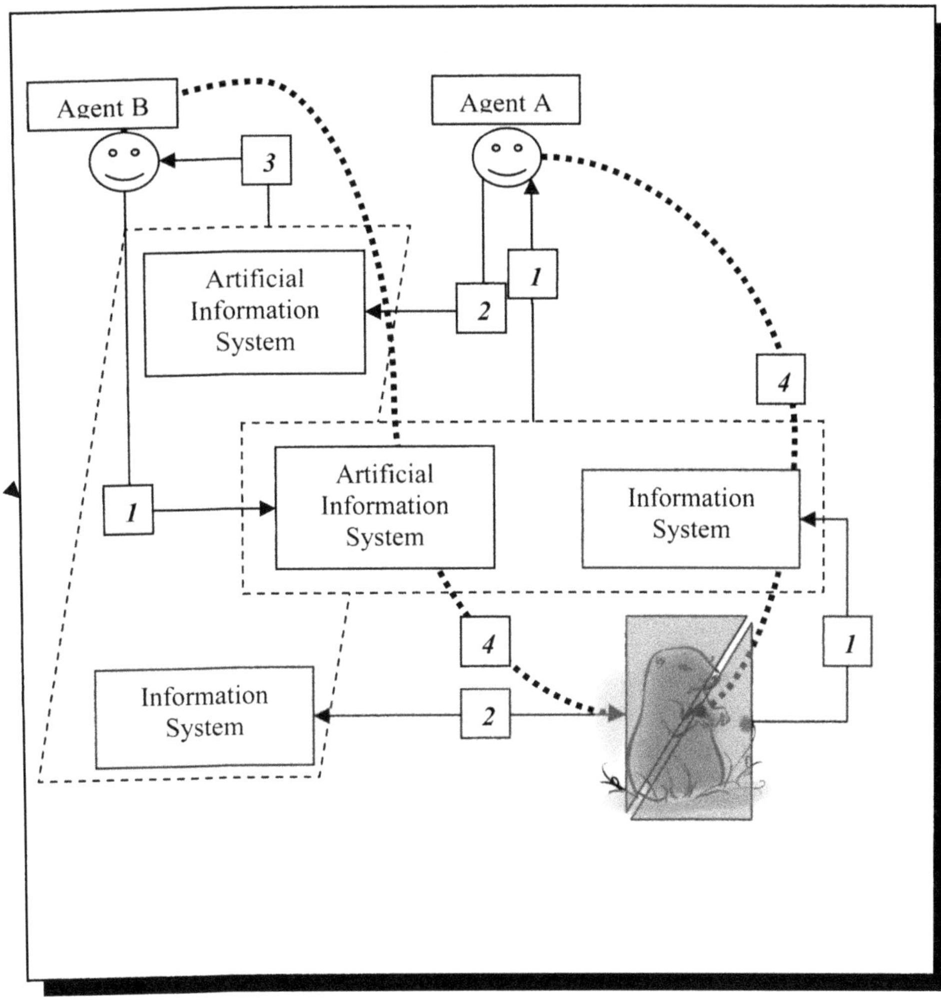

(4) *B* coordinates its actions according to the expected behavior of *A* inferred from (3). *A* and *B* act upon the reality.

Figure 5.1 Interaction between agents *A* and *B*, so that *B* can coordinate its actions with the actions of *A*. The success of the coordination will depend on how reliable *A* is to *B*.

The accuracy and breadth of the information systems involved in this process determine the predictability of the behavior of *A* as presented to *B*. The precision and completeness of these information systems determine how reliable *A* will be to *B*.

Our proposed notion of trust requires one additional concept, which is important to our general framework, namely the notion of *ideal agent*.

The ideal agent presents all attributes and predicates one specific agent considers praiseworthy to find in another agent. These attributes and predicates generate information systems and actions in accordance with the values inherent to them. The ideal agent is identified by an agent with the mythic notion of a super-hero, as conceptualized by that agent.

Notice that an agent must not necessarily wish to become an ideal agent, or to interact with ideal agents. The ideal agent deserves the utmost admiration of an agent, who for practical purposes may prefer to interact with less than ideal agents, and may limit its ambition also to less than the ideal agent.

Trust relates directly to the identification with the ideal agent. Agent *B* trusts agent *A* to the extent that it identifies *A* with its conceptualization of the ideal agent. The more *A* is perceived by *B* as a super-hero, the more *B* trusts *A*.

A super-hero may not necessarily have predictable behavior. Hence, trust does not imply reliability. Conversely, an agent whose behavior is wholly predictable may not necessarily be motivated by noble and beautiful qualities. Reliability and trust, therefore, are independent concepts.

What trust and reliability have in common, nevertheless, is that both are judged by an agent based on information systems which carry as components artificial information systems built by the agents that are being analyzed. An agent *B*, therefore, *believes* that an agent *A* is reliable and an agent *A'* is trustworthy.

The consequences of misjudging reliability are easy to characterize. Misplaced reliability amounts to actions of *B* based on expectations of actions of *A*, which are then frustrated. The consequences of misplaced reliability are breakdowns on coordinated actions.

The consequences of misjudging trust are a little more subtle. Misplaced trust amounts to *B* being frustrated about the intrinsic nature of *A'*, which was possibly perceived as holding noble features that, later, are revealed not to be part of the constitution of *A'*. The consequence of misplaced trust can be a breakdown on the whole perception of reality, since the information system which was in use to connect agent *B* with reality may then be deemed untrustworthy.

Reliability is thus an instrumental concept, useful for agents to coordinate their actions and interact with other for practical purposes, e.g. to reach goals and execute tasks.

Trust, in contrast, is not a guide for action of any sort. Trust is a self-fulfilled notion, a goal in itself instead of a tool to reach something else.

Considering that you, dear reader, assume the position of agent *B* in the examples and discussion above, who would you like to have around you? Agents on whom you rely or agents you trust?

Evidently, if an agent has goals to reach, plans to execute and tasks to develop, it will wish to have reliable agents around it with which to interact.

When would a trustworthy agent be desired, then? Our view is that the will to be in contact with trustworthy agents depends on a further notion, namely the notion of *self-trust*. The closer an agent sees itself as its own idealization of an agent, and the stronger is to an agent the goal to effectively become its idealized super-hero, the more this agent will desire to interact with those agents whom it can trust, i.e. who are noble representatives of its notion of super-hero.

The path to self-trust is what makes trustworthy agents desirable. And the way to ensure that self-trust will not be frustrated goes through self-reliability. The will to perfect self-trust and perfect self-reliability leads to the perception of an agent about itself that is unmediated by information systems. As will be developed in the chapters to follow, this corresponds precisely to our conceptualization of *self-knowledge.*

Who would you like to have around you? Agents on whom you rely or agents you trust?

Bibliographical notes

The Coordinated Attack Problem is a classical problem in theoretical computer science. A nice formal description of this problem and how to tackle it can be found in (Fagin et al, 1995).

Trust is a notion whose importance has grown substantially in recent years, in diverse areas such as the design of autonomous systems in artificial intelligence, management theory and economics. An interesting book on trust is (Solomon and Flores, 2003).

Intermission

It is an interesting fact that the etymological root of the words *jealous* and *zealous* is the same.

Dr. Agnelotti is very zealous of his back garden. He has always been so. Especially now, that the back garden is about to become a two lane road according to San Isidro director plan. Especially now, after we have decided to preserve the back garden for Giovanni.

We are working today in our mosaic. Hopefully, we will end up with a beautiful picture, made of certainly beautiful pieces, which are going to be fine representations of the memories, stories, feelings, emotions, shapes, colors, sounds, tastes, textures and scents of the Agnelotti back garden.

The back garden is full of meaning. Every corner holds a myriad of surprises. The observation and annotation of all details of the back garden is going to be an extensive job.

We decide to split the job in two halves, one for each of us. Dr. Agnelotti is going to start from the back wall – the ancient wall, close to the hennery and the tomatoes – and I am going to start from the front gate.

All things going well, we shall meet at the fig tree.

We may have some additional work to align our terminology, notation and the granularity of our representations. This can be left for later.

At least we are going to have a starting point. Something we can refine later, to reach some baseline quality standards that make us happy.

Dr. Agnelotti takes a notebook and a carefully sharpened pencil. I take the laptop and a digital camera.

We move to our corners of the back garden.

Dr. Agnelotti starts shouting from the back garden. He asks me whether he has already told me about when Gigino fell down the wall. I stop my work to listen. The back wall is almost three meters high.

Superman – the movie – had just been released. The whole family had gone to the movies on Sunday. Gigino truly enjoyed the movie.

After that, and for the following days, Gigino took a piece of red cloth and tied to his neck as a cape. He played all the time pretending he was Superman.

Mrs. Agnelotti anticipated the disaster. She commented one evening she hoped Gigino did not try to fly like Superman. Knowing Gigino as they did, Dr. Agnelotti and his wife smiled to each other and sighed.

They heard Gigino screaming aloud the movie theme song. Then silence. Then 'bump'. Then Gigino crying.

They rushed to the back garden. Gigino had climbed the back wall. It remains a mystery how he managed. Then he flew. Straight down to the floor. Three meters.

Almost a routine. Grab Gigino, check hastily the seriousness of the bruises, and rush to the Royal Infirmary.

I enjoy the story, as usual. I am sure there are a lot of fine details to be added to it. Humorous details, poetic details, fine pieces of this mosaic that explain more and more about the Agnelotti family.

However, we have work to do.

I prefer not to get too involved in the story this time. Dr. Agnelotti can write it down in his notebook using his beautifully sharpened pencil. We can read it later. It can be a good way to check whether we are being successful in our work.

I turn my back to Dr. Agnelotti and return to the front gate.

I take some pictures, and carefully tag them. I transfer the tag codes to the computer, and start writing notes.

Front gate. New one. Fifteen years old. Replaced heavy wooden one, beautifully carved by unknown artist. Cicilla knocked old gate with car. Dr. Agnelotti was reassuring. Cicilla did not resent the disaster. A good laugh for everybody.

Dr. Agnelotti can walk really quietly. I do not notice him leaving his post to peek over my shoulder.

"Do not forget to mention the delicacy of Cicilla when she said 'Oops! Sorry!', it is an important point in this story". I only notice Dr. Agnelotti when he interferes with my work suggesting the inclusion of detail that seemed fundamental to him. The suggested detail seems colorful to me, but I do not think it is fundamental.

The suggestion annoys me. I decide not to mention what Cicilla said when she left the car.

Dr. Agnelotti keeps waiting.

I keep staring at him and waiting for him to get back to his corner of the garden, so that I can continue with my work.

We turn our backs to each other. Dr. Agnelotti returns to the hennery. I turn to the olive tree.

Olive tree. Close to the front door of the house. Very beautiful. About two meters tall. Thirty one years old.

Dr. Agnelotti starts shouting from inside the hennery. He asks whether he has told me about when Cicilla was locked inside the hennery.

I resist.

This is not a good time.

We have work to do.

I shall not turn to listen to the story.

I shall concentrate on my work.

Mrs. Agnelotti shouts from the kitchen whether we would like a glass of orange juice.

I run to the kitchen. I would love to hear about when Cicilla was locked inside the hennery.

Cicilla was four years old. She loved the chicken. Her always delicate and smooth movements did not scare the chicken, so she could get really close to the chicken, to her delight.

It was one Sunday morning. They were having breakfast under the fig tree. Cicilla had finished her breakfast and started playing in the garden. She pretended she was a hen, and flocked with the chicken. The whole family laughed.

The door bell ringed. It was Mrs. Agnelotti's sister. They moved their attention from Cicilla to Mrs. Agnelotti's sister.

After a while, they moved to inside the house.

About two hours later, Mrs. Agnelotti's sister went away. They stayed inside the house. Mrs. Agnelotti went knitting, Dr. Agnelotti read the newspaper.

It took them another hour at least to realize that Cicilla had been too quiet for too long.

They started looking for Cicilla. No one could find her. They started feeling really scared.

After lengthy and suffering minutes, Mrs. Agnelotti remembered that Cicilla was playing with the chicken when her sister arrived.

They ran to the hennery.

Cicilla had gotten inside the hennery with the chicken. The door closed behind her. Cicilla was trapped inside the hennery.

The chicken had eggs to hatch, so they went to their nests and stayed quiet.

Cicilla was pretending she was a hen. She went close to the chicken and accommodated herself to hatch her imaginary eggs.

And then Cicilla fell asleep.

They found her sleeping peacefully together with the chicken.

Dr. Agnelotti took a picture. Mrs. Agnelotti found it and showed it to me. It was worth a prize.

We finish our glasses of juice and go back to our corners of the back garden.

Olive tree. Cicilla planted it. She was two years old when she did it.

Maybe I should add a brief statement about what Cicilla said when she knocked down the front gate.

"Gigino tried to eat an olive directly from the tree once. It tastes horrible, as you know. Gigino hated it."

Dr. Agnelotti is again peeking over my shoulder. I stare at him and wait for him to return to his side of the garden.

This is my side. My side.

Dr. Agnelotti stares back at me. His eyes look calm and patient, as always.

Dr. Agnelotti moves back to his side of the garden. I can continue my work on my half of the garden.

I see a scratch on the front wall, close to the gate. I ask Dr. Agnelotti about that scratch.

I prepare myself for another story. Would Gigino have bitten the wall when he was five? Would Cicilla have maneuvered the car in some unsuspected way and hit the wall? Would Dr. Agnelotti have marked the wall on purpose so that the family would always remember some momentous event?

Dr. Agnelotti comes to have a look at the scratch. He had never noticed it. He shouts to Mrs. Agnelotti whether they still have the paint stored on the basement. Mrs. Agnelotti shouts back, "what paint?" Dr. Agnelotti explains it is the paint they used to paint the front wall. There is a scratch on the wall, he explains to her.

I feel a little disappointed at first.

Dr. Agnelotti looks at things in the back garden and sees events and memories I cannot see.

I look at objects in the back garden. They are all fresh and new to me. I can notice details Dr. Agnelotti is not aware of.

Dr. Agnelotti's view of the back garden discloses mysteries and secrets my eyes cannot see.

My view of the back garden reveals mysteries and secrets even Dr. Agnelotti may not know.

It would help if Dr. Agnelotti did his work quietly, and made sure he would have had a fair amount of work done by the end of the day.

It would help if Dr. Agnelotti stopped peeking over my shoulder and let me do my work quietly, so that I can have a fair amount of work done by the end of the day.

I continue my observations. Every detail counts. I try to use the drawing program that is installed in the laptop. I want to add a pictorial description of the scratch in the wall. I add some dimensions. I take pictures.

I feel like a policeman.

I bow to the power of words.

Drawings are no good. Pictures are acceptable, but drawings may not be included. I am not skilled at drawing. And the program we have in the laptop is not a very good one.

Words are good. With words we loose the unspeakable. On the other hand, with words we can talk about all of the "speakable".

Words convey attitudes. Words convey actions. Word is action.

Not every word, of course. But with words we can bring about the sense of action. And action itself.

"In the beginning was the Word..."

I describe the scratch with words. I include the fact that it was discovered by me. I include date and time. I want to catalog the scratch.

I include the reaction of Dr. Agnelotti and of Mrs. Agnelotti. I start to write some notes about the unveiled possibilities of that scratch in the wall.

I realize that perhaps I should be more succinct. Especially about things I have little information about. I turn back to the olive tree to see if there is something else to be added about it.

I notice that Dr. Agnelotti is strangely quiet. It has been nearly fifteen minutes since he last shouted about something or came to peek on my work.

I look at him. He is there, sitting on the floor close to the tomatoes. He seems to observe attentively the tomatoes. I notice he is taking notes every now and again.

Productivity, at last.

I imagine Dr. Agnelotti is writing some story related to the tomatoes. I guess it must be a good one. I look at the tomatoes and see the most beautiful tomatoes in the world. This is all I see.

What can Dr. Agnelotti see and feel when looking at those tomatoes? Which tomatoes is he observing? The ones in front of him? The ones he remembers were there when something nice and funny occurred with them?

I wish I could know.

I wish Dr. Agnelotti would come and tell me.

I consider going there to ask. Or perhaps to peek over his shoulder.

I resist the temptation.

I could find the scratch on the front wall. I would probably be able to find something about the tomatoes, too. Something new to everybody.

The tomatoes are on Dr. Agnelotti's side of the garden.

I should better concentrate on the olive tree.

I count the branches, I estimate their angle with respect to the floor, I observe the size and color of the leaves. I notice that one branch seems to differ from all the others. It seems ill. There may be something wrong with it.

I consider shouting to Dr. Agnelotti to inform about my finding. I look at him first. He continues to look concentrated in the tomatoes. The story must be good.

I conclude that it would be a bad idea to interrupt him. I can talk about the sick branch of the olive tree later.

The roots of the olive tree are quite close to the two steps at the entrance of the house. The front door has beautiful engravings. It is made of dark wood. It separates the back garden from the internal corridor that gives way to the dining room.

We have decided to focus on the back garden. I shall not cross the front door.

The engravings have some sort of religious motif. It seems like the three Wise Kings crossing the desert towards a small hut. There is a flowerpot at each side of the door. I do not know the name of the plant; it bears some large beautiful purple flowers.

I annotate everything I see in the laptop. Every now and again I take a picture, and add a reference to the picture tag displayed in the camera in the text.

Dr. Agnelotti continues silent. It has been a long time now. I look at him. He has moved closer to the hennery again. I notice that he is not holding the notebook anymore.

Dr. Agnelotti is picking eggs.

A few days ago I would have felt really annoyed with this situation. Here we are, doing our work. Each of us has his responsibilities. We must work as a team; otherwise the project will not be accomplished in time. And Dr. Agnelotti interrupts his work to pick eggs.

I realize I do not feel angry or annoyed this time. It seems the right thing to do at this moment. All things have their time. There is a rhythm to everything. It is time for Dr. Agnelotti to pick eggs now.

This is the way things must be. Otherwise, we would have no reason to wish to preserve the back garden. Dr. Agnelotti must stop his taking notes to pick the eggs, or he would not have notes to take.

This is part of the back garden we so much want to preserve.

I take a picture. I write a note in the laptop. A separate piece of the mosaic, it will be. An important piece of the mosaic nonetheless.

When I look again Dr. Agnelotti is no longer there. I turn again to the front door. I hear Dr. Agnelotti speaking with his wife. I look again and Dr. Agnelotti is back to his post at the tomatoes.

Birds.

Some birds sit on top of the olive tree. I turn very quietly and take a picture of them. Beautiful small birds. Very noisy birds, too.

I realize we are not recording sounds. I decide to leave it this way. I am going to describe the sounds in the text. I bow to the power of words.

Some birds sit on top of the olive tree and I write about them. I count them. About six birds. All of the same species. Small, grey feathers, a little white spot on the chest. They are very noisy. They apparently do not fear humans. They are gone.

A dog barks in the street. We can hear the sounds from here. Cars, dogs, people walking and talking. An airplane passes by; we can hear a very distant roar.

I look above the front gate. I can see the sky. Blue and clear. We can see the mountains today. We cannot see them when it is not so clear. Dr. Agnelotti can estimate the humidity of the air by looking towards the mountains. Today is a very dry day, for example.

Mrs. Agnelotti shouts from the kitchen door. She complains we have been too quiet for too long. She asks if we would like a glass of fresh water.

Dr. Agnelotti and I race towards the kitchen door. Mrs. Agnelotti laughs. Dr. Agnelotti, of course, wins. Quite fit he is, Dr. Agnelotti.

Dr. Agnelotti asks his wife if she remembers when Gigino decided to make a scarecrow for the tomatoes. Mrs. Agnelotti laughs and says of course she remembers. Gigino took her best brooms and cut the broomsticks to build the structure of the scarecrow. The head was an old bucket – at least he chose the old bucket for that, she says – and the clothes were made of old newspapers.

Gigino has always been skilful with craftwork. He managed to build a human sized scarecrow that looked very much like a person from a distance. He did that with very scarce material. Mrs. Agnelotti realized she did not have her brooms anymore after the scarecrow was already installed in the middle of the tomatoes.

A scarecrow was not necessary, of course. There was always enough movement in the back garden to keep the birds away from the tomatoes. But it added the likings of a farm to the garden, and everybody enjoyed it. Dr. Agnelotti had to buy new brooms, and they had to ask Gigino not to take things without asking for permission first, but that was all.

Mrs. Agnelotti warns us that within less than one hour lunch will be ready. We offer her some help, she laughs and winks almost imperceptibly at Dr. Agnelotti, and says we should better continue our work and be productive. She will want to know how many pages we have prepared in the evening. Depending on the answer, we may not be allowed to go to the ice cream shop.

We rush back to the garden like two kids. We three laugh pleasantly.

The door bell did not ring this morning. I decide to talk about it nevertheless. There is a door bell. It sounds like an ordinary door bell. It is very old, yet it works quite properly.

Maybe I can elaborate on the door bell later. I am sure there is a better way to explain how it sounds.

I am sure Dr. Agnelotti can tell stories about the door bell. I am sure he will be reminded of so many people who have used it. Some people use it very regularly, almost every day, like Ms. Conchetta for example.

Some rather unusual people have also used the door bell. I recall for example the story they told me, about when the Mayor came to ask if the restoration in the street was still causing floods at the lower part of the back garden. He came in for a cup of tea and ended up staying for dinner. The morning after the restoration in the street was over and the external side of the front wall was repainted.

There are many more details to this story. I hope Dr. Agnelotti remembers to write about them.

Dr. Agnelotti, however, is responsible for the other side of the garden. The door bell is my responsibility.

We may need to rethink things a little bit.

It is lunch time. We can discuss this after lunch.

Mrs. Agnelotti has prepared aubergines stuffed with minced meat. It is delicious. We eat that and drink a glass of wine each. A true banquet.

Today we celebrate our very productive morning. Mrs. Agnelotti warns us, however, that we have not earned our ice cream of today yet. We promise to work harder in the afternoon.

Before we restart our cataloguing work, I suggest to Dr. Agnelotti that we discuss a little about a methodological issue that occurred to me.

Dr. Agnelotti says he suspects he knows what this issue is about. He asks if it has to do with the scratch on the front wall.

I answer that in fact it does. It came to me when I was reflecting about the door bell, but indeed it is related to the scratch on the front wall.

We both nod and smile to each other. We suspect that we have arrived at similar conclusions.

I suggest that perhaps we have divided the work inappropriately between us two. Our first go to it was to have half of the garden completely described by Dr. Agnelotti, and the other half completely described by me. It seems to me, however, that I could never record things about the memories and sensations of the back garden the way Dr. Agnelotti could. For the simple reason that I have not lived the experiences that have generated those memories and sensations. Those have to be written by Dr. Agnelotti.

On the other hand, my perceptions as of an outsider reveal details about the appearances and perceptible relations among objects in the back garden that could pass unnoticed to someone who has lived there for so many years.

Our views complement each other.

Our annotations should complement each other.

A combination of our annotations shall convey a better compilation about the back garden to Giovanni than just the juxtaposition of the annotations.

I say that and start looking for my empty cup. It is a matter of seconds until Dr. Agnelotti suggests that I should have a cup of coffee.

Dr. Agnelotti agrees with my suggestion. No coffee offered this time.

I smile proudly.

Dr. Agnelotti suggests that, for the sake of having a discipline for the work, we continue in our posts. We can build our partial views of our respective halves of the back garden, and then continue over the other half. This shall avoid that we bump on each other all the time, as well as get distracted by each other all the time.

Once each of us covers the whole back garden, we can share our impressions and combine our notes.

We will be building half pieces for the mosaic. Each half piece prepared by me will be combined with a half piece prepared by Dr. Agnelotti.

Seems like a good plan to me. Very concrete and direct. Just the way I like plans.

We move to our posts. I feel confident that Dr. Agnelotti will trust me on my work now. I shall be doing something I can do better than anyone here about this back garden. I am not going to invade the work of Dr. Agnelotti. He can do his work better than anyone at all.

I understand why he was peeking over my shoulder all the time before. It is just natural he could not trust me on doing *his* work.

So why is he coming this direction then?

Dr. Agnelotti suggests that we make a little experiment. Just to see how the end result of our work will look like. This can be helpful for each of us, so that we have a guideline for our work. It can also be an incentive for us to continue working, since we can have the taste of what the end result of our work will look like.

I like very much the suggestion.

We must choose one topic to work as an experiment.

I suggest that we work on the front door. I was taking notes about the front door before lunch, so I have a point to start. I have not heard any story related to the front door yet, so frankly I would like to listen to any. I am sure there must be many, as there seems to be a multitude of stories under each stone on the ground in this back garden.

Dr. Agnelotti agrees. The front door it shall be. Our initial experiment. BackGarden version 1.0 is slowly taking shape.

I can take the notes in the laptop. I suggest we indicate who is contributing with what in the notes. Dr. Agnelotti agrees.

We shall use different fonts. Dr. Agnelotti's contributions shall be in italics, mine shall be in normal font.

> The door looks heavy. It is made of dark wood. The wood used to make this door must be very dense. Yet, it opens easily. When it moves, it makes a funny squeak.
>
> The engravings are very beautiful. Apparently, they represent the three Wise Kings crossing the desert. *This door was installed here not too long after Dr. Agnelotti and his family moved to the house. There was a plain door here before. Not*

very nice, the previous door. And it was old and a little rotten in some points.

The whole family helped to choose the new door. Gigino of course did not help much, he was too young for that. Cicilla was also very young, but she expressed her tastes quite well already.

The engravings are very well done. Some details are really nice. The kings are traveling on top of camels, and the camels are engraved to the finest detail. The scene depicts them in perspective, as if they were moving away from us, towards the middle of the door. We can see some details of the garments of the last king in the row, who is larger than the others because of the effect of perspective.

It is certainly not the result of mass production. Some artist must have spent quite some time carving this image in this door.

Craftwork is typical of San Isidro. There are many artists here, and they work really well. Twice a year there is a crafts fair at the central square, which I have been told is really nice.

The whole family went to the carpentry workshop. Mrs. Agnelotti carried Gigino. Cicilla walked, holding hands with her father.

Mrs. Agnelotti preferred a simple door. All they needed was a door in good state. A plain door would do perfectly well. They still had some furniture to buy. Mrs. Agnelotti preferred to save some money on the door and spend it on a new dining table later.

Dr. Agnelotti was very fond of those beautiful doors, engraved all over. He has always appreciated this kind of craftwork. He thought that the front door was the greeting card of a house and a family, a beautiful front door was like a warm greeting to incoming guests.

Engraved doors were of course quite more expensive than the plain ones.

The doors were displayed in large frames. They looked like gigantic books that could be browsed by the customers.

Mrs. Agnelotti could not turn the pages herself, because she was carrying Gigino. Dr. Agnelotti had to turn the pages of the Book of Engraved Doors for his appreciation, and then turn the pages of the Book of Plain Doors for Mrs. Agnelotti.

Cicilla moved with him from one display to the other, holding hands with her father.

At some point, Dr. Agnelotti let Cicilla go. A moment later, Cicilla had disappeared.

Dr. Agnelotti got really scared. He was afraid that Cicilla could get in between the doors, and the doors could crush her. The doors were quite heavy, and perhaps one of the displays could be a little unleveled.

They looked around, and Cicilla was not around the gigantic Books of Doors. They felt relieved. Now they needed to find her.

Suddenly they heard her voice. It sounded joyful. It sounded as if it were coming from the other room.

They went there, and there they found them both. Cicilla and their future front door.

Cicilla was pointing to the door. She was smiling and muttering words no one could understand.

The carpenter explained that he was carving that door for Saint Agustin Church. He had made a mistake, however, and the door was too small. He was going to destroy the carving and restart it all over. It was probably going to become a plain door.

Dr. Agnelotti winked at Mrs. Agnelotti. She smiled.

Dr. Agnelotti argued that, if that door was going to become a plain door, then it had to be sold by the price of a plain door.

The carpenter smiled. He did not wait Dr. Agnelotti finish his argument. He said it was OK, if Dr. Agnelotti did not care having a religious motif on his front door then he could take it.

By the price of a plain door.

This was how Cicilla found the front door, Dr. Agnelotti got his engraved door and Mrs. Agnelotti got her new dining table.

The Agnelotti family is very religious. In a nice and well informed way. It is beautiful to see how they all cultivate their spirituality, by following the rites while at the same time reading and studying a lot about all dimensions of Religion in human History.

The three Wise Kings at the front door fit perfectly well with their religiosity. It announces to those coming to visit them that in that house Religion is taken seriously.

The Agnelotti family is also marvelously practically minded. One would not find a grain of rice in that house that would not have a concrete practical reason to be there.

That front door – a fully engraved door bought by the price of a plain door, in such way that it was a great deal also for the carpenter – is a living symbol of that collective practical mind.

Giovanni will be able to learn many things from that front door. It deserves some more pictures.

Pictures taken.

The front door welcomes the visitors. The front door secures the house. *The front door tells a little about who the Agnelotti are.* The front door holds important and interesting information about the regional craftwork around San Isidro. *The front door integrates with the back garden. It is part of the back garden. It is part of the Agnelotti.* The front door is the frontier. It separates the back garden from the house. And the house is a whole different universe. *There are no different universes. There is only one universe. And the Agnelotti belong to it.*

What is the Agnelotti back garden? Where does it start, and where does it end?

The back garden is limitless.

Evening is approaching. It is getting dark. Soon it will be dinner time. Dr. Agnelotti and I are blissfully happy.

We have production to show to Mrs. Agnelotti.

I carefully save the files and make backup copies.

We have earned our day.

We have earned our right to the ice cream shop.

BackGarden version 1.0. First piece of the mosaic is done.

Chapter 6

Ontological Reasoning

"That's a good one! We have a subject who exists but doesn't realize he does and there's my paladin who thinks he exists but actually doesn't. They'd make a great pair, let me tell you!"

Italo Calvino

By *ontological reasoning* we mean collaborative problem solving in which a communication space is agreed upon by the participating agents prior to the communication actually taking place.

Hence, ontological reasoning implies collaborative problem solving in which several simplifying assumptions take place. The most evident one is that the shared communication space is precisely defined and accepted *a priori* by all participating agents. Therefore, it is not built interactively as a result of ongoing communication, and it is known in advance by the participating agents, who can access it with utmost accuracy.

When the agents belonging to a group accept a pre-defined communication space to interact, these agents also accept that their communication is going to be *grounded* by that communication space. In other words, the price to be paid for by the employment of a pre-defined communication space is that the connections among the participating agents are constrained by the pre-established communication space.

A pre-defined communication space does not determine internal meaning for information structures perceived by each agent. It only determines the co-occurrence of certain structures perceivable by different agents, namely those structures that are mapped to structures in the communication space.

When we introduced the notion of communication space, we stressed that it was an artifice to convey some notions in a more palatable way. A communication space was an inexistent information system, which would hopefully be envisaged as the end result of an infinite process when things went well in a dialogue. This artifice has been given existence through pre-established simplifications of the possible interactions among agents, embodied as *ontologies and their instances*.

Ontology

The word *ontology* is an ancient word, whose meaning can be traced back to the writings of Aristotle. Recently, however, it has been revamped to stand for a technical concept useful to design and implement computer software. It is to this revamped meaning of ontology that we refer in this chapter.

An ontology is a partially specified communication space, which characterizes a collection of communication spaces through completion of its specification. In purely software engineering terms, it corresponds to an abstract class which captures the notion of an abstraction of a communication space. The concrete instances of this abstract class are specific communication spaces derived from the ontology, just as a concrete class or an object is derived from an abstract class in object oriented programming.

When used in computer applications, an ontology can be implemented as a partially specified data structure plus a set of rules specifying how it can be completed. The data structure can be as sophisticated as desired, including components such as static data structures, dynamic data structures, functions and procedures.

An ontology, therefore, comprises the specification of a collection of concrete communication spaces, namely the communication spaces that can be derived from it.

Ontologies are pre-fabricated, off-the-shelf components to build communication spaces. A group of agents can build – or accept – a communication space from a given ontology, and then by mutual agreement accept this communication space as the means through which they are going to interact. By doing so, they of course agree on a set of structural constraints upon which they are going to build their individual perceptions about a portion of the reality, about themselves and about each other.

We are carefully avoiding any assumption about the internal meaning that can be given by each agent to any part of its private information systems, including the artificial information systems built for the purpose of communication. We are only assuming that, if agents engage into a productive dialogue, the artificial information systems they build for this purpose will all map onto the instantiated ontology they have agreed to employ. We also assume that their shared goal for collaborative problem solving is fully represented in this instantiated ontology.

The appropriate semantic notion to utilize for an ontology, therefore, is akin to the notion of *operational semantics* found in theoretical computer science and in the specification, design and analysis of programming languages. More specifically, the semantics of an ontology (*viz.* of any instance of an ontology) are the syntactic constraints imposed by its data structures and completion rules (plus possibly some additional rules to capture syntactic interdependencies between portions of these data structures and rules). An ontology can be used by any group of agents and in any situation in which the desired interaction among these agents is such that the constraints imposed by this operational semantics does not bar the agents from representing important facets of their desired interactions, and it will be useful when these constraints match the communication space those agents would build by means of their interactions.

For purposes of didactic *presentation* of an ontology, its author can present it through examples, resorting to specific domains of application, i.e. portions of reality whose internal structure and organization are such that the proposed ontology can be instantiated as a fine communication space for them. The components of the ontology, however, do not have as their *meaning* the components of the example in use. Rather, they *correspond* to the appropriate portion of reality syntactically and structurally.

An instantiated ontology is a communication space used for collaborative problem solving. It is therefore an artificial information system used for communication. As such, it contains perceptions of an agent about a particular facet of reality, about itself and about the other agents with whom it interacts. When agents collectively agree to employ an instantiated ontology, they assume that their views about reality and about each other all match with what is proposed in that instantiated ontology. It should not be surprising that these assumptions break up eventually during the interactions. An important and complex issue related to ontologies and their applications, as a consequence, is how to monitor and to maintain these communication spaces.

As a concrete example, let us consider an agreed upon syntax to represent geographical information using maps. One such syntax is an ontology in the sense depicted above. Any instance of this ontology is a concrete map of a region, containing the information that can be represented and communicated using the syntactical elements of the ontology.

A specific map is an off-the-shelf communication space. An agent can draw and annotate a map employing the graphical syntax proposed in the ontology, to function as a communication space between this agent and a second agent who is going to consult the map and learn about a certain region through it. Notice that, by consulting the map, the agent can give up – at least provisionally – direct contact with the region that is represented in the map. These two agents – heretofore named respectively *cartographer* and *reader* – interact via the map, thus giving up their direct contact with reality on behalf of the artificial information system whose *operational* interpretation is shared by both (Figure 6.1).

The cartographer replaces her contact with reality by her personal interpretation of that contact with reality, or else with a portion of her personal interpretation of her contact with reality that can be conveyed through the particular form of representation comprised by a map. Her choices about what to represent and transmit through the map take into account a stereotyped map reader, to whom the map is built.

The reader, on his turn, has goals to be reached that justify his interest in information. Some additional reasons must justify his interest in obtaining information from a map, instead of direct contact with reality – cost effectiveness, time limitations, accreditation rules that consider a standardized artificial information system more reliable than subjective personal experience, etc. – in which case the reader shall consult the map, thus replacing the information system he could build by actually going to a certain place by an information system whose contents refer to the cartographer, to what the cartographer has proposed to be the relevant portions of the reality to be represented in a map, and to the cartographer's perception of the reader, his goals, capabilities and plans.

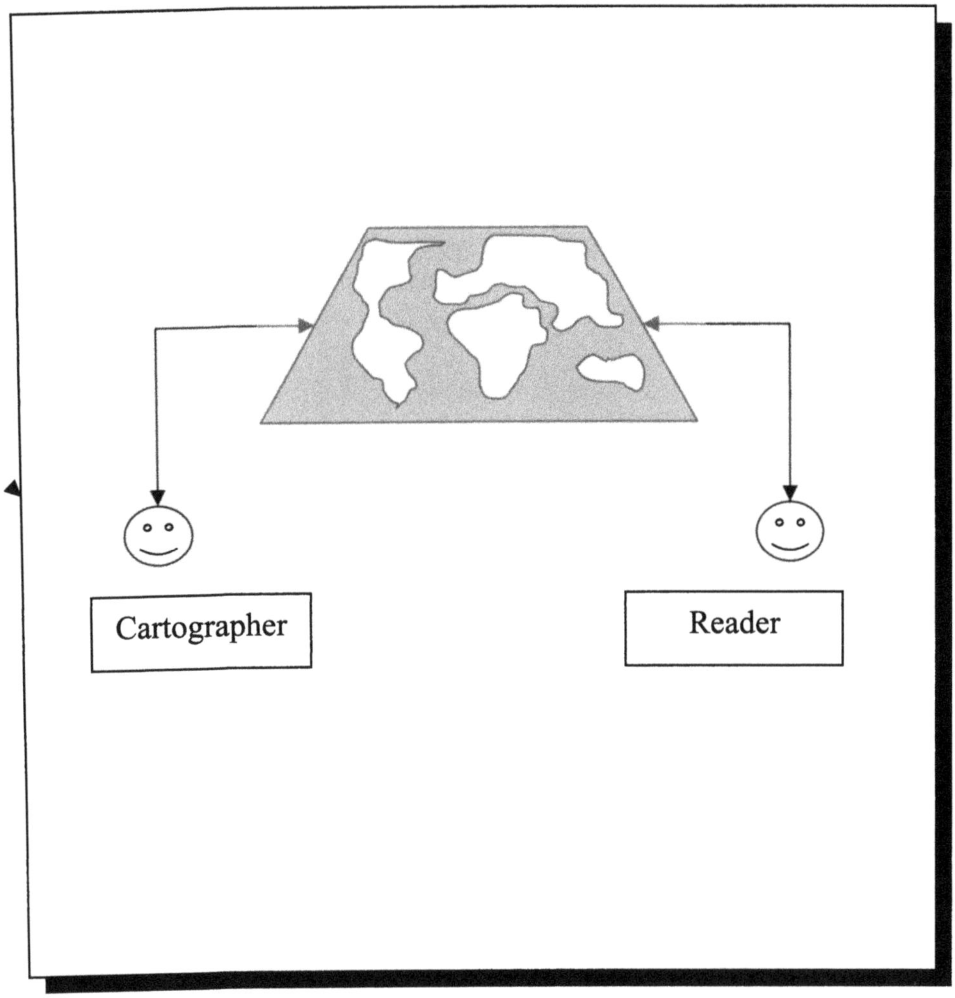

Figure 6.1. Maps as instantiated ontologies for the interaction between a cartographer and a map reader

Maps can be considered from other perspectives. Let us consider, for example, the case of a map being used by an agent to ask another agent about directions to reach a certain destination. In this case we have at least three agents engaged in interaction – the agent who is asking for directions, who we will call the *tourist*, the agent who is providing directions through the map, who we will call the *local citizen* and the *cartographer* who drew the map in the first place (Figure 6.2).

The cartographer makes use of an ontology of geographical representations to build a map. This map is then adopted by the tourist and the local citizen so that they can interact to determine the directions to follow. The three agents are engaged in a process of collaborative problem solving, making use of an instance of a pre-established and agreed upon ontology – the map – which is assumed as a communication space. These three agents have different goals, which are nevertheless synchronized. Evidently, the interaction protocol among these three agents is not symmetrical, in the sense that their participation in the dialogue varies depending on the specific stereotyped roles each of them assumes (cartographer, tourist and local citizen).

The parallel between this example and the instant messaging system considered in Chapter 3 should not be difficult to be drawn. Cartographer and system designer have similar roles in both examples, namely to build an artificial information system based on direct perception of a selected portion of reality – a place in the case of the cartographer and a mode of interaction in the case of the system designer in charge of developing an instant messaging system; the pair tourist / local citizen and the users of the instant messaging system, on their turn, are the ones who accept the initial perceptions and corresponding proposed artificial information systems of, respectively, the cartographer and the system designer, and engage into direct interaction already guided by those artificial information systems.

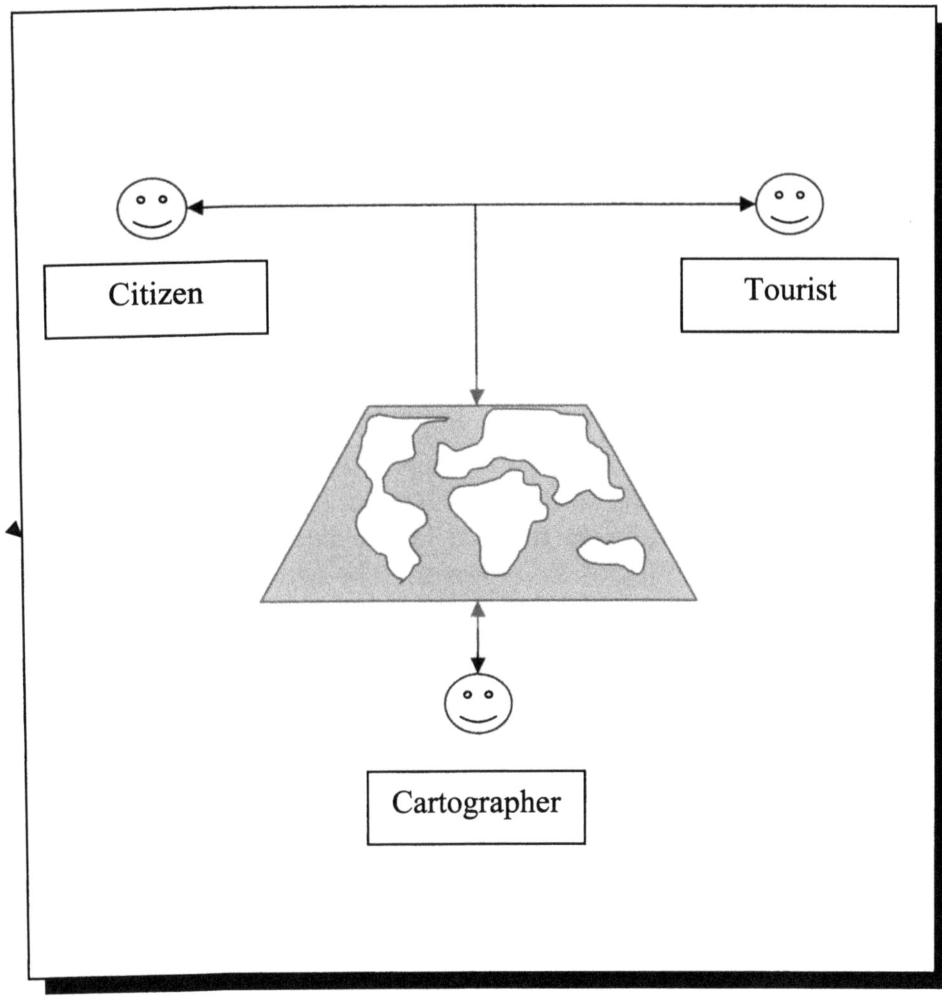

Figure 6.2. Maps as instantiated ontologies for the interaction among a cartographer, a tourist and a local citizen

In both cases, any analysis of the communication spaces being used by these agents is incomplete if it does not take into account all participating agents in the dialogue and their specific participation in the interaction process – based on their perceptions of the reality, of themselves and of each other, their goals, their selected languages to represent the reality and to interact with each other, and the technological constraints imposed by the available media, tools and devices that mediate the interaction.

Patterns of interaction

Ontologies are therefore computational products, to be designed, built and sold by specialized companies. Indeed, many companies have been created in recent years devoted to this business, some of them very successful. Ontologies have proven to be useful patterns to build communication spaces for special purpose interactions which are by design simplified and stereotyped, to serve specific shared goals and enable collaborative problem solving based on standardized and simple protocols.

In very broad terms, these protocols for interaction can be organized in a few patterns, which suggest certain structures for the ontologies and thus drive their design and construction. The existing ontologies, as well as their corresponding communication spaces, can be classified according to these patterns, and each pattern of interaction contains its own set of attributes to assess the quality of a specific product.

The patterns of interaction can be organized considering three dimensions:

1. The number of participating agents,
2. Whether the participating agents know each other, and
3. Whether the interaction occurs in real time.

Any interaction assumes the participation of at least two agents, one for the role of sender of artificial information systems and another for the role of receiving agent. The number of participating agents refers to how many agents assume the role of senders and how many agents assume the role of receiving agents in an interaction. Following the tradition of software and database design, we group the possibilities in the cases of "one" and "more than one" agents for each role. Hence, we have four possibilities:

1. *1-1*: one sender agent and one receiving agent participating in the interaction. This is the precise case of a dialogue involving exactly two agents.

2. *1-N*: one sender agent and many receiving agents participating in the interaction. This is the situation for example we can observe in a lecture, in which the lecturer is the sender agent and the audience is the group of many receiving agents.

3. *N-1*: many sender agents and one receiving agent participating in the interaction. This situation, which in some sense is the reverse situation of the previous one, can be found for example in a group therapy session, in which the patients are all sender agents and the therapist is a receiving agent for them all at once.

4. *N-N*: many sender agents and many receiving agents participating in the interaction. This is the situation that can be observed in a variety of social interactions, such as a company departmental meeting in which all participating agents take turns as sender and receiving agents, or a football match, in which the players interact also by taking turns as sender and receiving agents of information systems.

Clearly, and by analyzing just this dimension of characterization of patterns of interaction, our classification is very rough. The class of *N-N* interactions could be further refined, for example, to consider interactions in which some agents specialize in being senders or receiving agents, thus characterizing more precise patterns of interaction that might suggest more specific desired attributes to be found in an ontology.

Likewise, we could refine the notion of "knowing each other", thus producing more refined predicates to be found in specific ontologies. Here, we consider an extremely rough notion of knowledge, perhaps that would satisfy only a policeman trying to identify a suspect: an agent *A* knows another agent *B* if *A* can identify *B* out of a crowd of agents. We have then two possibilities:

1. The agents engaged in an interaction *know* each other, i.e. they know with whom they are dealing. This is the case, for example, in a company departmental meeting in which the participants all know each other.

2. The agents engaged in an interaction *do not know* each other, i.e. they are interacting anonymously with each other. This is the case, for example, when an author writes a book and an audience of readers later buys and read that book.

As for the interaction being or not in real time, we have essentially two possibilities:

1. Real time interaction is that sort of interaction that requires simultaneous attention of all participating agents. This is the case, for example, of a phone call, a company departmental meeting and a lecture.

2. Non real time interaction, on the other hand, permits that agents interact asynchronously. This is the case, for example, of electronic mail messages, letters and books.

The combination of these three dimensions gives us sixteen patterns of interaction. Each pattern of interaction induces certain patterns for the structuring of communication spaces, which drive the design and construction of specific ontologies.

Bibliographical notes

Among the recent bibliographical references that characterize the notion of ontology as employed in the realms of Information Technology, one that is very frequently cited is the one found in (Gruber, 1993), which states that "an ontology is a specification of a conceptualization".

One evident problem related to ontologies is the strong requirement they implicitly demand, that agents accept the rules of interaction proposed in that third-party product in order to interact. More recently, some computational products have been proposed to address this problem. Essentially, these products are virtual environments that enable groups of agents to collectively and dynamically build the ontology they intend to use for their interactions, thus adding to the ontologies novel degrees of flexibility and adaptability that largely enrich them as tools for disciplined and purpose-driven interaction among groups of agents. The interested reader can find some information about these recent developments in ontological engineering in (Gruber, 2005).

The operational semantics of programming languages refers to the meaning of information structures ascribed by their computational behavior, instead of meaning dependent upon reference to external objects and domains. This is a mature and mathematically precise field of theoretical computer science. A fine reference to the semantics of programming languages in general, which includes the presentation of operational semantics, is (Mitchell, 1996).

Intermission

We are very enthusiastic about our recent progress in the cataloguing of the Agnelotti back garden. We have found means to preserve important facets of the back garden for Giovanni. This way he will be able to learn quite a few things about the place where his family lived for so many years. He will also learn many things about his family, based on the experiences that we register as connected to the back garden. Eventually, Giovanni will learn a lot of things about himself, by learning things about his ancestors.

Our work is going to contribute for Giovanni to know himself.

It is breakfast time. Here we are, as usual, under the fig tree. I look up to this magnificent tree and I still find hard to believe that it is simply going to be cut down in a few days. Consequences like this seem to be the ones we always find in director plans. Some plans fail, others partially succeed, none fully succeeds. The common factor about director plans is that they cut down trees and move squares and gardens to different places and force people to change their way of living.

OK maybe I am not very happy about this specific director plan. There may be other common factors connecting all director plans in the history of western civilization.

People make plans all the time. We plan our route to the market, we plan our parents' fiftieth wedding birthday party, and we plan a business project that is going to involve many people in different continents and a vast amount of money. Politicians make plans all the time. Urban planners – no wonder they are called this way – make plans all the time.

Some of these plans are called *director plans*.

All director plans are plans. Not all plans are director plans. Some director plans are successful. Others are disastrous.

Plans can be classified. There are universal properties that identify many abstract objects as *plans*. Add some other properties, and from plans you detach those that are director plans. Add yet more properties and you have the successful director plans.

And so on.

The Agnelotti back garden is a representative of a general class, namely the class of back gardens. Back gardens are in turn a special case of *gardens*.

In between back gardens in general and the Agnelotti back garden there are some further levels of classification. Some back gardens are irrelevant. Others have the power to describe to the minute detail the history and personality of a family.

The Agnelotti back garden, of course, falls into this latter category.

It occurred to us that some descriptions and stories that we have written and some pictures that we have taken are related specifically to this back garden. Others, however, are about any back garden of some relevance. Others are about any back garden whatsoever, and there are even stories and observations that apply to any garden.

When talking about the Agnelotti family, sometimes we are so specific that our observations and stories can only apply to this family. Some other times, however, we could be talking about San Isidro. Other times we could be talking about being human and about mankind.

I understand it is not easy at all to identify when we are talking about anecdotes that relate specifically to the Agnelotti and when we are talking about human nature.

If we could at least identify some points in our notes that refer more generally to human relations, families in general and mankind, we could publish these points so that they could be accessed by a greater public. They can be useful for more people, who may be interested in understanding a bit more about how they relate to each other and why certain events are so important while others are truly irrelevant.

Before we continue with our annotations of specific points and stories of the Agnelotti back garden, we decide to exercise a little on generalizations. Our hope is that we will be able to produce generic abstract descriptions that can be somehow "detailed" to concrete cases in order to produce specific memories of different families.

If we are successful, one day a farmer in Botswana may be able to use our pre-fabricated abstract back garden to build a description of his property to his grandson.

If we are *truly* successful, one day a grocer in Bangalore may be able to use our pre-fabricated abstract structures to build a database of his business, so that he can die in peace knowing that his descendants will have the means to understand the value and organization of each facet of his commercial activities.

The Agnelotti back garden has a front gate. A front gate separates the property from the environment that surrounds the property. A front gate must have moving parts, so that it can be opened and closed. Some front gates must also have additional devices so that it can be locked.

When the front gate is shut, people outside the garden cannot see inside. Some people inside still can see a little of the outside, provided that they are tall enough. This is true of some front gates. Other front gates are sufficiently high so that no one inside the garden can see outside, and vice-versa.

There are still the front gates that are so small that they let everybody see each other. These are, so to speak, symbolic front gates.

The category of gates can, therefore, be refined as sight cluttering front gates, symbolic front gates and those half way through. The front gates that fall into the "half way through" category depend on the user to be reclassified as belonging to one of the former categories. The two first categories are mutually exclusive.

Front gates are very useful. As we hope our generic descriptions are going to be.

The Agnelotti front gate evokes stories about the Agnelotti family, which describe this family, how its members relate to each other and how these relationships were built along the years.

Front gates sometimes are like this.

Other front gates are silent. They do not tell stories at all. They just sit there, being opened and closed depending on circumstances. They are a little more than geometric boundaries that determine the shape of a piece of land that belongs to someone.

Front gates, as gates that they are, also determine – or at least influence – the path people use to walk in and out of a garden. They are therefore conveyors of certain types of behavior. Gates constrain and influence the activities of individuals who inhabit the space for which they work as boundaries.

Front gates are installed according to plans. Depending on the breadth and depth of the plans, one could even assert about certain front gates that they are the result of director plans.

The relationship between Dr. Agnelotti and Cicilla was strengthened through this front gate. Actually, through the previous implementation of a front gate that preceded this one. Geometrically speaking, they are equivalent, so we do not loose much when we refer to both gates – the old one that was smashed into pieces by Cicilla and the present one – as a generic *Agnelotti front gate*.

The Agnelotti front gate can reveal a lot about how Dr. Agnelotti educated his daughter. It was also a fundamental element that helped Dr. Agnelotti educate his daughter.

This was however a very specific event that exposed some very specific facts relating Cicilla and Dr. Agnelotti.

The facts *per se* can be referred to as particular instances of the whole relationship that existed – and still exists – between Cicilla and Dr. Agnelotti.

It does not take a professional psychologist to notice that this relationship is a very healthy and constructive relationship between a daughter and a father.

Which is a particular element in the category of healthy relationships between daughters and fathers.

Which is a special case in the category of relationships between daughters and fathers.

From all human relations, we can identify those relating people from the same family. From those, we can select the relationships between daughter and father. From those, we can leave aside the pathological ones and consider the healthy relationships.

As with high and short front gates – and those half way that are high to some people and short to others – human relations can be divided into evidently healthy ones, evidently destructive and pathological ones, and those that stay in between. Indeed, the majority of human relations fall in between.

The relationship between Cicilla and Dr. Agnelotti is an exceptional case. The Agnelotti family is an exceptional family. Their relations are completely healthy. The longer I stay here with them, the more I admire this family.

I have noticed very subtle remarks from Mrs. Agnelotti implying that things have not always been like this. They have had their bad times, too. They have learned from experience. They have made mistakes.

I believe Mrs. Agnelotti. Wisdom cannot be bought from the drugstore. One has to build it. And quite often – nearly always, I would say – the path towards wisdom is steep.

The Agnelotti have built collectively a beautiful way to relate with each other, with the others and with everything in the world. They are simply great, in a very unique way.

Their greatness is one of a kind. A very unique greatness among all possibilities for greatness. A distinguished element belonging to the set of greatnesses, if we can talk about such set.

People can be great in a variety of ways. Attitudes can be great. Feelings can be great. Relationships can be great. Greatness is a quality. It is applicable to a great variety of entities.

Greatness can also be qualified in many ways. There is a specific sort of greatness that applies to a specific sort of relationships, namely family relationships involving a father and a daughter. From that sort we identify this very special element, which explains how Cicilla relates with her father.

From and through this relationship, and as a consequence of its greatness, we understand how and why Cicilla is who she is.

We also understand how and why Dr. Agnelotti is who he is. The chronology of events must be subverted sometimes if we want to go one step further in the understanding of human relations.

Cicilla educated Dr. Agnelotti very well indeed.

We find on planet Earth many great families. They have always existed. We find on this planet many healthy relationships. These relationships have always occurred located in time and space. They have left their footprints wherever and whenever they have occurred. Human relations leave traces in objects. One can infer the relations from the objects.

Just like archeologists sometimes do.

I must say I am always intrigued by how these fantastic scientists can look at a minuscule piece of a broken pot and infer from that a complete religious ritual that took place thousands of years ago.

It takes a lot of studying to do that. It takes a lot of work, and a lot of patience to be able to do that. The ritual is imprinted in the piece of broken pot nonetheless.

You look at the front gate at the Agnelotti back garden and you should be able to see the other front gate there. The new one reveals the old one. The old one – the one that no longer exists – reveals the greatness of the Agnelotti family, here represented by how Cicilla and Dr. Agnelotti relate to each other.

The front gate is a representative of the objects that bear imprinted in their essence the footprints of human relationships, past and present. The front gate is a particular specimen of a species.

We appreciate the reach and the ambition of what we are doing. We are taking notes about the human race. We are looking through this back garden to the infinitude of possibilities of human relations.

Human relations are not shapeless. They are not abstract. They have shapes and forms. They occur in places and times. You can see them. You can touch them. You can smell them.

Human relations leave behind echoes of their existence.

Look at the Agnelotti and you are looking at all people. Look at their back garden and you are looking at all places where these people have been.

Consider a tree. Any tree. Not necessarily a particular tree.

A tree has stories to tell. Look at it attentively and you may be able to hear them. It may have people having breakfast under it. It may have birds building nests on its branches. It may have kids falling from their top and breaking arms and bringing parents hushing to infirmaries with their kids on their arms.

The fig tree is a tree. It has all these stories to tell. It talks about the Agnelotti, more specifically.

Trees talk about people. The fig tree talks about the Agnelotti. The Agnelotti are people. The fig tree is a tree. A special case of a general class of events.

Does the general class really exist? Does it make sense to talk about the general class?

Or would the class be just a bunch of individual cases?

Is there a connection?

Is there a unifying feature?

Would the Agnelotti back garden encompass the whole planet?

Is it everywhere?

Can it be destroyed at all?

Where is this garden?

What are we talking about? Dr. Agnelotti, the whole Agnelotti family, all Mediterranean people who have lived here since the Pleistocene? Human race?

Am I included in the back garden?

Can I escape from it?

Can these questions be answered at all?

Can these sketches of grand theories about the human race be of any use to the farmer in Botswana? Can they help make survive the grocery in Bangalore?

Can any of these be of any use at all to Giovanni?

Giovanni who?

Oh, yes, I was nearly forgetting. Giovanni, son of Cicilla, grandson of Dr. Agnelotti. The real culprit. We started all this because of him.

OK perhaps we have changed subject a little. We wanted to record the Agnelotti back garden for Giovanni. This was the initial plan.

The reason to do that was because the back garden reputedly carried much more information than what a simple observation of its appearance could tell. The back garden was a representative of the history of the Agnelotti family itself.

It came out to be that the real interest was in preserving the history of the Agnelotti family for the future generations – Giovanni included.

The back garden gave shape and existence to this history. A significant part of this history had occurred in the back garden. By preserving the back garden we would be preserving the history of the Agnelotti, in a very genuine way.

The Agnelotti are at the same time very unique and very universal. Certain predicates only apply to the Agnelotti. I sometimes think we should invent some words to describe these people, as none of the existing ones really captures the originality of the Agnelotti. Other characteristics of the Agnelotti, on the other hand, make them proud and fine representatives of their people. Of the people of San Isidro. Of the Mediterranean millenary culture. Of human race.

Their back garden, it must be said, is an ordinary back garden. You would find many back gardens very similar to this one, even here in San Isidro.

Yet, it is so unique. It bears the history of a family.

Furthermore, as if it were a stage where a play had been presented showing the history of the world, it bears the history of a lot of people. Of all people, depending on how you look at it.

Certain things are worth preserving to tell to Giovanni. Other things should be kept secret. He will have to learn them by his own means. He can check later how those things were already there, imprinted in his family's back garden.

A fig tree is a relatively large tree, whose leaves are big and rough. It bears figs.

What does this say about the fig tree below which we have breakfast every morning?

It tells a lot.

It tells nearly nothing.

It is a starting point.

We must show Giovanni a picture of a fig tree. A generic one. From an encyclopedia, for example. Then we must show him a picture of *this* fig tree. Then the figs. Then some figs that have come from this tree.

Giovanni must taste at least one fig. It will help a lot if he can hug a fig tree. Any one. Preferably one tree about the same size of this one.

Then he can see the picture of this fig tree.

Then he can be told the stories about Gigino. About the birds. About the cardigan.

We can take the pictures especially for Giovanni. We can write down the stories especially for Giovanni.

No one can prevent other people to enjoy these stories and see these pictures, though.

Indeed, no one should try to prevent other people to do so.

Giovanni will be able to understand things better if he can learn about the general things as well. It will help if he can read the encyclopedia. See the pictures. Taste a fig. Any fig.

There is no separation.

You learn about the universals, and you are learning about the individuals. You study the individuals, and you are learning about the universals.

You study the universals and you reach the individuals. You lean on the individuals and you find yourself. You look at yourself and you see the Universe.

You look at the back garden and you see a generic back garden. You study back gardens and you see human relations. Universals. Individuals. Yourself.

You look at yourself. You see back gardens. The Agnelotti back garden. The Agnelotti family. Dr. Agnelotti.

Quite puzzling indeed, this back garden.

Where does it start? Where does it end?

Director plans have distinguishing features of their own. You know when you have found one.

They move people from here to there. They create roads where there used to be gardens and houses.

They make people think about their past. About their present. About their future.

Director plans pretend they are moving things from a place to the other. They actually create the opportunity for people to appreciate that things cannot be moved, for the simple reason that each single object is actually present everywhere.

Director plans pretend they are pushing history forward. Director plans pretend they are at the service of progress. They actually create the opportunity for people to appreciate that things cannot be forgotten, or left behind. For the simple reason that everything is always occurring. Who could ever expect the illusion of time being unmasked by something as simple as director plan?

Director plans are apparently disastrous. Some director plans are apparently very useful.

All director plans are fantastic opportunities for people to re-evaluate their lives, understand better how they relate with their environment, with each other, and with themselves.

Director plans create the opportunity for us to understand what it really means to be human. Director plans create the opportunity for us to identify what is worth preserving from our traditions, culture, and history.

Director plans create the opportunity for us to appreciate the inevitability of the preservation of our traditions, culture, and history. They provoke us to look at those things at which we look every day, and see in them those things we have never seen before. They show us that what we see must not be seen through such and such object.

Thanks to the San Isidro director plan, and to Giovanni, we have learned a lot today.

I believe we can start choosing what flavors of ice cream we shall have today.

From my own evaluation, we have already earned it.

Time for a cup of coffee. Before I have it offered to me.

Chapter 7

Uncertainty

From describing the past, from the present which seized my hand in its excited gasp, here I am, O future, now mounting the crupper of your horse. What new pennants wilt thou unfurl before me from towers of cities not yet founded? What rivers of devastation set flowing over castles and gardens I have loved? What unforeseeable golden ages art thou preparing – ill-mastered, indomitable harbinger of treasures dearly paid for, my kingdom to be conquered, the future...

Italo Calvino

Ontologies provide agents with the ability to start communication very efficiently and with unprecedented immediacy. Through an agreed upon ontology, a group of agents can build a tailor made communication space or accept an off-the-shelf one, and immediately start coordinated action or collaborative problem solving.

There is a price to be paid for all this efficiency, though, since there is no way to ensure that the communication will be always successful: the interacting agents can only take for granted that they have managed to build artificial information systems from which certain pieces could be extracted and mapped into a third-party communication space. Every agent *assumes* that the other agents have done the same, and moreover that their mappings from artificial information systems to the communication space are compatible with each other, so that the pieces of information that flow between agents preserve a natural notion of *coherence*.

These assumptions can only be checked empirically and *a posteriori*, i.e. agents interact and then perform coordinated action based on the expectations of each others' actions built by their interactions. After the actions take effect, the agents must go through a new round of interactions to check whether their expectations have been fulfilled and their interactions have been effective. Positive reinforcement leads to further interactions and coordinated actions, through which a group of agents builds collectively an increasingly refined sense of mutual reliability.

From an individual agent's perspective the perceived events are interpreted through information systems. There cannot be absolute certainty that the agent's perceptions correspond to the reality. The agent acts based on these perceptions, and these actions include information flow. Information flow induces in other agents a sensation of shared understanding, which leads to a system behavior when we consider the system of all interacting agents. Even if there is no absolute certainty that agents understand each other, their interaction generates a certain behavior of the community as a whole. Regardless of how successful each agent considers the interactions with other agents, its perceptions and its actions which should be coordinated with the actions of other agents, the simple acknowledgment of the existence of other agents to interact, and the expectations about these interactions that are raised by a past history of interactions and a resulting sense of trust, changes the behavior of each individual agent, and consequently of the community of agents.

Trust is, therefore, a creative attitude, in the sense that it causes the appearance of a communitarian behavior, and this is true even when we have partial trust. Indeed, in the vast majority of cases, agents will only partially trust each other. Only in exceptional cases we will have an agent completely trusting or completely distrusting another agent.

Controlled uncertainty

Our focus in this book is the set of interactions among agents which are based on or at least involve the exchange of information. We are therefore interested in interactions which involve the exchange of artificial information systems, which are tailor made for each interaction and such that their contents filter some portions of a perceived reality out of the information to be communicated, as well as combine together some other portions of the perceived reality as indistinguishable representatives of idealized events that comprise a simplified *model* of reality.

Agent interactions based on information exchange are, therefore, inherently imprecise and incomplete. As discussed in previous chapters, it is precisely thanks to the imprecision and to the incompleteness that information systems are useful, since it is through imprecision and incompleteness that the communication becomes efficient and that the information systems can be recorded and stored as symbol structures.

In other words, agent interactions based on information exchange are inherently pervaded by uncertainty. The utility of such interactions to reach goals and execute plans based on specific purposes depends on how efficient the communication processes are, but also on how each agent manages to control the uncertainty that originates from the information based interactions.

The incomplete and imprecise rendition of the reality that is available to an agent when it replaces direct contact with reality by the access to reality mediated by information systems and other agents, guides and drives the actions of that agent. Actions, however, interact directly with the reality. The missing bits that separate the reality from the information systems that effectively guide the behavior of the agent must be completed by the agent, based on hypotheses, presuppositions, inferences and past experiences of the agent. An agent controls the uncertainty that is inherent to dealing with the world through information by refining its skills to complete the information systems that are available to it through the set of capabilities listed above.

Consider any cards game with incomplete information, bets and guesses. Consider poker, for example. A poker player has a very precise goal to reach – win the game – which is based on a very precise general purpose – to earn as much as possible of the currency being bet in the game (money, tokens, etc.) – and based on which he builds and follows plans and strategies to play well.

The poker player, however, does not know everything he should know in order to optimize his performance. He knows the cards that are on his hand, but he does not know the cards that are on the hands of his opponents, nor the sequence of cards that are still in the deck. Furthermore, the player does not know everything about the behavior, goals, purposes and plans of his opponents.

The missing information is a pre-requisite for the poker player to decide what action to execute next. The player must therefore complete his information. In order to do so, the player must assume that the other players are going to behave based on some pre-specified patterns – for example, that all players are going to obey the rules of the game that is being played – that the deck of cards is a standard deck for the game being played, and so on. The player must also infer as much as possible about the hands of the other players, by memorizing what cards have already been used, by observing the expressions of the opponents, etc. Eventually, the player will have to guess some missing information in order to complete his picture of the reality.

A good player minimizes the amount of information he must guess. By doing so, the agent completes the information system that has been made available to him by reliable means, resorting to the least reliable alternative – plain guess – as little as possible.

A *very good* player, on top of that, always bears in mind that the available information is inherently incomplete and imprecise. His actions and plans are guided by this very important additional information, so that for example he never bets all his currency on a single hand.

Every information-mediated interaction among agents is like a poker game. Some interactions are potentially more reliable than others, to the extent that the corresponding relevant information systems can be completed more effectively by the participating agents through reliable means.

The extent to which a potentially reliable interaction among agents becomes an effectively reliable interaction depends not only on the nature of the information systems being exchanged, but also on the skills of the participating agents to exploit all facets of the reality that are being conveyed via the exchanged information systems.

Since every information-mediated interaction is pervaded with uncertainty, the actions that result from such interactions have also some inherent risk, related to the possibility of some other agent or some event in the reality not behaving as expected. The expected behavior of other agents and the reality is constructed based on the available information and on the gaps that are filled in through hypotheses, presuppositions, inferences, memory and guesses.

This risk should be minimized – indeed, this is what is expected from *rational agents* – but cannot be eliminated, since it is part of the process.

An agent who decides to eliminate the risk of communication breakdowns and consequent misguided actions from its interactions with other agents severely constrains its possibilities of interaction. The vast majority of information-mediated interactions among agents require the risk that is natural to acting based on the expectation of the behavior of other agents.

Consider, for example, electronic commerce. An agent offers some product on the Internet, accompanied by a description of its attributes, as well as the attributes of the company that is offering the product (delivery time, warranty, etc.). Based on this artificial information system, a second agent decides to buy the product. Most often than not, the buyer knows very little about the seller, and has to fill in the gaps through analogy with similar past experiences and guesses. If the buyer decides to buy goods over the Internet only when she knows everything about the seller, it is likely that the buyer never buys anything at all through the Internet.

Similarly, assuming that the buyer decides to buy a product through the Internet, the seller must fill in gaps about her, e.g. assuming that the credit card information that is being sent to him is reliable. If a seller does not assume some risks to sell goods through the Internet, it is likely that he never manages to complete any transaction.

Electronic commerce, therefore, is a direct result of agents accepting the inherent risks of their information-mediated interactions, assuming that these risks will result from some lawful patterns of behavior and interaction. These lawful patterns of behavior and interaction are characterized by the structure of the procedures and protocols used in the interactions. In other words, electronic commerce exists because the participating agents accept that the risk involved in their interactions follows certain rules, which are stated in the communication spaces that are used by the participating agents for that specific pattern of interactions. The communication spaces must therefore predate the interactions, i.e. they must be established in advance, so that the agents can subscribe to the relevant communication space to interact.

The standard way to encode pre-established communication spaces are ontologies. By using pre-established communication spaces, agents can engage into novel patterns of interaction, whose most distinguishing feature is the possibility of controlling the risk of breakdowns.

Trust, confidentiality and privacy

As can be concluded from the previous section, reliability and trust are pre-requisites for many of the patterns of interactions that can be observed in our modern, Technology founded society. This perspective for the design of interactions among agents is not so frequently analyzed, though, nor used to guide the design, implementation and utilization of systems for information-mediated interactions.

Building computer applications for people to interact in the age of the Internet can be understood as building digital environments to enable the construction of reliability and trust. The idealized utilization of digital technology is to create novel possibilities so that people can gather new perceptions of the reality and of each other, share these perceptions in more refined and accessible communication spaces, and therefore interact more effectively, reducing the risks of communication and interaction breakdowns.

Ideally, this should lead to an increasing interest in understanding each other and reality, and sharing this understanding through whatever means are available. This would include self understanding, which then would lead to increasing self reliability, self trust and ultimately the quest for increasing trust of each other.

Our purposefully optimistic view of the paths we are treading is that we shall gradually approximate this idealized situation, if for no other reason simply because this shall prove to be the economically most effective means of interaction.

People trust each other to the extent that they believe that they are dealing with approximations of their conceptualizations of a super-hero. The sense of trust depends on the availability of refined information for an agent about the other agents – to evaluate how close these agents are to its conceptualization of the super-hero – and about itself – among other reasons, to have a clear access to its personal *model* of a super-hero.

This construction of the ideal path for the evolution of digital technology could suggest that eventually, through information systems, people will open themselves completely to a universal communication space, to enable a complete sense of trust in each other and of self trust. There would be no place for confidentiality or for privacy in this setting, which does not quite seem like a fine ideal to be pursued.

The problem with this over-generalization of the notion of shared understanding is that it precludes the possibility of choice from the agents interacting through communication spaces. An agent either connects to this holistic community or stays out of it.

In this section we argue that this must not be necessarily the case. We can have our idealized setting and still have room for confidentiality and for privacy.

It should be recalled that the point here is information-mediated interaction, i.e. that sort of interaction into which a group of agents chooses to engage, given that each of these agents has a goal to be reached. No matter how ambitious or full of grandeur a goal can be, it is still a goal to be reached, that will cease to exist once it is reached and therefore that is ephemeral. These agents have needs to be satisfied, and it is because of these needs that they provisionally flock together and interact with each other.

It should also be recalled that the agents we are considering in this work are human beings, who are notoriously complex and multifaceted beings. It is natural to consider that an agent can have more than one goal to achieve, and even that these goals can be conflicting or compete for resources. Hence, it is natural to consider that an agent can wish to engage into interactions with a group of agents to achieve a goal, and with another group of agents to achieve a second goal. The agent may choose not to inform its peers from one group about its interactions with the other group, if for no other reason simply to be economical in the construction of artificial information systems, and to avoid clogging those information systems with pieces of information that may be irrelevant to carry on a certain sort of interactions.

Hence, a communication space that joins together a group of agents willing to interact, and such that the risk of breakdown in communications and interactions is minimized, is perfectly compatible with a notion of privacy, in which the information that is added to the communication space by each individual agent is carefully selected so that some facets of the agent (as it is in reality) are preserved from sharing.

Likewise, if an agent participates in more than one group of interacting agents, it can choose not to disclose to one group its participation in other groups, with no harm to any interaction with any group. A natural notion of confidentiality is therefore compatible with our notion of ideal interactions: an agent is entitled to preserve as confidential its interactions with one group of agents, when interacting with a second group of agents.

Bibliographical notes

Agent interactions based on imperfect information is a field of investigation by itself, particularly relevant to fields such as Economics, Sociology, Marketing and Logistics. This field of investigation is usually called *Game Theory*. In this chapter we have avoided this name – at least until this point – since we have not even scratched the underlying mathematical models and theories that are normally employed in this field. Essentially, our goal was to use the terminology developed in the previous chapters to discuss the problems that are relevant to information-mediated interactions. Some readers may nevertheless find this field of investigation interesting in itself – and interesting definitely it is. An interesting and inspiring introductory text on Game Theory can be found in (Dutta, 1999). A more concise introduction to this subject can be found in (Bicchieri, 2004).

Intermission

There is a general characterization of gardens, of which the Agnelotti back garden is a particular case.

There is a general characterization of human relations, of which families are a particular case.

There is a general characterization of families, of which the Agnelotti is a particular case.

There is a general characterization of observers, of which Dr. Agnelotti and I are particular cases.

There is a general characterization of observations, of which the notes we have been taking are particular cases.

There is a general characterization of how observers and observations can relate to each other, of which our experiences in the last few days are particular cases.

Dr. Agnelotti and I have started the construction of general theories which, when instantiated to particular cases, shall build our desired depiction of the Agnelotti back garden. That one which is about to be transformed into a two lane road. All that for the consumption and satisfaction of Giovanni. When he gets to the proper age, of course. Giovanni-at-the-age-of-five, for example.

We believe we can describe certain universals, to which special features can be added in order to build particular instances. Among all possible particular instances, we find the one that interests us. The starting point and the desired end point of this journey.

Among all possible particular instances, one should be able to find also many other possibilities, which can be useful for many other people. Starting and end points of so many journeys, of so many journeymen we now meet along our way. A farmer in Botswana. A grocer in Bangalore. And so many others.

All of us talking about the same universals. About particular cases of the same experiences. About being human. About being alive.

Are we really?

Could it be that we deceive each other, living an illusion of universal communion and understanding, when in fact we are alone, enslaved in our perspectives and impressions that cannot be shared at all?

I say that a specific leaf of the fig tree is dark green and large. I say it is beautiful. I say it feels rough when I touch it.

What does the farmer in Botswana understand when I say this?

What does the grocer in Bangalore understand when I tell him that the fig tree has dark leaves?

What will Giovanni understand?

Is all our work worth anything at all?

What happens when we communicate with each other?

What gets through? What is blocked on the way? What is caged inside the speaker?

I say "red". What do you see? What do you feel?

I say "Giovanni". What do you feel? What do *I* feel? What does Dr. Agnelotti feel?

How can we know?

Yet, we talk. We write. We see, taste, smell, touch and hear.

At least we believe we do.

How many back gardens are there out there? One? Two? Many? None?

How many back gardens are there inside me? One? Many? None?

How many back gardens are there inside Dr. Agnelotti?

How do these back gardens relate to each other?

I come from a region where fig trees do not grow. It has to do with the climate. Figs are quite expensive where I live. They have to be imported. Or at least used to have. I have read recently that some farmers in a remote corner of the country have succeeded in planting fig trees. Quite an achievement.

Fig trees are very common in the San Isidro area. They can be found in many gardens.

I like them.

Dr. Agnelotti says "fig tree" and I point to a certain direction. Dr. Agnelotti nods, approving the direction I am pointing. I am pointing to the correct object. According to him.

We have a common understanding of the fig tree, after all. Something has been created among Dr. Agnelotti, the fig tree and myself. Some sort of bond.

I cannot *really* know what Dr. Agnelotti sees when he looks at the direction where the fig tree is. I cannot *really* know what he feels. I know more about this now than a few days ago. I know the stories. I have been sitting under this tree every day.

Dr. Agnelotti probably can guess a few things of how I feel about the fig tree. He cannot *really* know what I see when I look at the fig tree. He cannot know what I feel about the fig tree.

And of course neither of us makes the faintest idea of what the fig tree thinks about ourselves.

Yet, we say "fig tree" and we look at the same tree. When I leave San Isidro, I will be able to identify a fig tree when I see one.

I have learned a few things about fig trees in general.

I have learned a few special things about *this* very special fig tree.

Communication is not completely useless after all. We do exchange some sort of patterns that can induce behavior on others, which is somehow predictable.

Dr. Agnelotti has his mysteries. Among all his mysteries, there are his impressions about the fig tree. They are enshrined within him; they live in that sacred cloister I see in front of him just now, which turns to me if I call his name – "Dr. Agnelotti".

His impressions about the fig tree are secret; no one can ever unveil this secret.

I also have my private mysteries. Among them, my impressions about the fig tree. Not that I would not like to tell them to Dr. Agnelotti. Indeed I would. I simply do not know how.

Yet, not everything Dr. Agnelotti knows and feels about the fig tree is secret. Not everything I know and feel about the fig tree is secret. Certain things can be shared. Certain things can be communicated.

My views about the fig tree are influenced by what Dr. Agnelotti communicates to me. My secret impressions have been partially built in cooperation with him. And perhaps I have also contributed to build his views about the fig tree.

Communication can be good. Communication can be useful.

The fig tree itself has told me many things about itself. It has been good to communicate with the fig tree.

By recording these messages to be communicated to Giovanni, we are creating opportunities for him. He will hopefully be able to build impressions about many things based on these messages. About his family, about where they lived, about where they live, about himself. About the world and human beings.

What exactly he will learn from these messages we do not know. We cannot know. These are going to be his mysteries.

Dr. Agnelotti says "front gate" and I point to the right direction. Mentally, I also point to a wooden gate I have never seen, to Cicilla as she was years ago, to the "Oops! Sorry!" I did not hear, to the archetypal relationship between fathers and daughters and how this relation manifested itself as the love Dr. Agnelotti feels for Cicilla, which in turn manifested itself in his reaction to that situation.

Dr. Agnelotti says "Giovanni" and I mentally point to his relation with his grandson, to the whole web of relations that links together this wonderful family. I also point to the picture above the piano. This is the image of Giovanni I have.

Maybe I should have confessed earlier. It is a good moment, anyway. Better said than kept secret.

I have never met Giovanni.

The last time I met Cicilla was on her wedding.

We have not met after that.

I do have pictures of Giovanni. Cicilla sent them to me by the Internet, and of course Dr. Agnelotti and Mrs. Agnelotti have told me many things about his first grandson.

I have never met him, though.

There is a possibility that Giovanni reads these notes and looks at these pictures one day, and we never meet.

BackGarden version 1.0. Impressions about the Agnelotti family and how they interacted with a particular space. As recorded by a complete stranger.

What is Giovanni going to know about me when he accesses BackGarden version 1.0?

Does it matter at all?

Considering things from this perspective, it seems a little bit presumptuous that we wished our notes to be useful to the farmer in Botswana.

What do we know about the farmer in Botswana?

What can he know about ourselves, or San Isidro, or Giovanni, or the reason we started all this?

He cannot know anything at all about this, of course.

Do they have fig trees in Botswana?

Perhaps that farmer can learn one or two things about fig trees. After all, it is written in a few places that its leaves are rough.

Perhaps we could write a few more things about how they can be used to shine the cutlery. He may even find some tree whose leaves have a similar texture. The farmer's wife shall be surprised when she arrives back home in the evening and finds her cutlery shining beautifully.

Perhaps an archeologist-inventor-mad-scientist finds and decodes our messages three thousand years from now, and is inspired by our notes to create a new product to shine the techno-trumpets of the Fifty-First Century Jazz Band.

What exactly are we writing about anyway? How many back gardens do exist?

Giovanni will learn about Gigino and Cicilla and his grandparents. Giovanni will learn to love and to admire them all.

Giovanni will learn about himself. He will learn to love himself.

Giovanni will learn about San Isidro. Ms. Conchetta, Mr. Fialho, his lovely aunt who passed by every morning.

Giovanni will learn important lessons about human relations. Ethics. Character. Principles.

Giovanni will learn a few things about me. I few a little embarrassed about this, I wish I could be simply a helping hand here. I cannot hide, though, even if I wanted to.

The farmer in Botswana can learn about fig trees. A scholar in Cambridge can learn about family relations in the Mediterranean traditional villages. The archeologist of the future can learn about culture and technology in our century.

We never tell the whole story to each listener. Or else, each listener never hears the whole story from us. The story is here.

The whole story.

BackGarden version 1.0.

I believe not even Dr. Agnelotti sees the whole story. There are subtleties and lines in between lines in between lines that perhaps not even us – the writers – are able to see.

Perhaps the archeologist of the future will see. He will learn things about ourselves, and about Dr. Agnelotti family, that we cannot see just now.

The "whole story", however, as we have already come to accept – with some reluctance, true, but also with resignation – is not the *whole* story of the Agnelotti back garden.

We loose some stuff on the way. We are carrying water in a leaky bucket. This is communication.

Other people add their stuff, though.

The illusion of life.

The illusion of time.

The archeologist of the future speaks to us.

We can hear him.

His bucket is severely leaky, though.

The stars up there.

We have not talked about the stars yet.

They look beautiful, as seen from the Agnelotti back garden in the evening.

They are not there.

They are messages from the past.

We look at them.

Our appreciation of their beauty does not reach them.

Not now, at least.

We are talking to the future. We are receiving messages from the past.

The illusion of time.

The illusion of life.

All things go really fine while we understand each other.

Dr. Agnelotti speaks "fig tree" and I point to the right direction.

Dr. Agnelotti speaks "basil". I look around. All bushes look the same to me. Dr. Agnelotti laughs.

I point to a beautiful bush at the corner, close to the wall.

Mrs. Agnelotti arrives with three glasses of water. They two burst into laughter.

That is parsley. So obvious.

The back garden cried "parsley". I heard "basil".

How often is this going to happen in our notes?

How often has the garden spoken to us *X* and we have written *Y*? How often has Dr. Agnelotti explained *X* to me and I have written *Z*?

How do *really* Cicilla and Dr. Agnelotti relate to each other?

Perhaps I have misunderstood everything? What does "Oops! Sorry!" really mean?

Can there be a very strange religion in central China, based on the belief that fig trees are the embodiment of God on Earth? Can our manifestations of love and respect for this tree be considered utterly sacrilegious to them?

Are we going to seal the fate of Giovanni with BackGarden version 1.0? Perhaps an army of fanatics is going to chase him forever because of our insolent statements about Mother Fig Tree.

How can we ever know?

Should we stop doing this work, to avoid the risk of misunderstanding?

How can we communicate without this risk?

We cannot.

We simply cannot.

Indeed, talk to anyone, talk anything you like. We cannot tell much about the result. One thing we can assure you, though. You *will* be misunderstood. At some point, at some deep level, somehow, somewhere, we can assure you. You will be misunderstood.

We know exactly what we mean by that. We imagine you also have a very precise understanding of what can be meant by that.

We can assure you that what we mean is not what you have understood.

See? We have already been misunderstood.

A risk worth taking, nevertheless.

We write about the back garden. We are misunderstanding important points about the back garden. We are missing many points. We may be adding points that were not even there.

There is much about the back garden in what we are writing, though. A distorted back garden, we concede to that. We are not so bad writers and photographers, though. C'mon. It is the Agnelotti back garden alright. It is impressions about the Agnelotti family that will amuse Giovanni. He will learn a few things here and there about *the* back garden.

Take no risks, and you are alone.

Take no risks, and you will not meet the love of your life.

Dr. Agnelotti met Mrs. Agnelotti. They were both young and inexperienced. They did not know each other. They talked to each other.

They still did not know each other. Yet, they built pictures of each other that made them look attractive to each other.

They got married.

They made mistakes.

They discussed a lot.

They felt sad at times.

They had Cicilla and then Gigino.

Sometimes they were too rigid. Other times they were too soft.

They made mistakes.

They learned. They still do.

They talked, and talked, and talked.

They used words, gestures, smells and noises and all.

They talked through their pores.

They talk a lot, the Agnelotti. They talk all the time.

They very generously talk to me.

I certainly misunderstand a lot of what they say. I certainly misunderstand their back garden.

They all seem to talk about the same thing. The *one* same thing. All the time we have been talking about one same thing.

The Agnelotti back garden. Where does it start, and where does it end?

The Agnelotti family. Where do they live? Who are they?

Who am I, and why am I here?

Who is Giovanni? Why do we want to speak with him? What do we want to speak to him? What would we like him to know?

What is he going to know?

We cannot know…

What is this whole thing about?

What are we talking about?

What is this message for? What is the message about?

We are talking about friendship.

Chapter 8

Knowledge Sharing

A page is good only when we turn it and find life urging along, confusing every page in the book. The pen rushes on, urged by the same joy that makes me course the open road. A chapter started when one does not know which tale to tell is like a corner turned on leaving a convent, when one might come face to face with a dragon, a Saracen gang, an enchanted isle or a new love.

Italo Calvino

Man, in the pursuit of knowledge, strives for understanding. The quest for integration and unity thus goes through distancing from that reality with which Man wishes to integrate.

We have already discussed in Chapter 4 the etymology of the words *knowledge* and *understanding*. We now add to the discussion the word *science*, which comes from the Latin verb *scire*, which stands for *to learn by distinction, classification and separation*. This verb probably originates from another Latin verb, *scindere*, which stands for *to cut and divide*.

The root of the scientific method, at least in its most accepted characterizations, is the classification of observable phenomena and the utilization of this classification to build predictive models for other phenomena similar to the observed ones.

Using our terminology, the work of the scientist is:

1. To observe the reality with well trained eyes capable of perceiving information systems not perceived before, most commonly than not biased by previously existing communication spaces which relate to the observed portion of the reality and guide the perceptions.

2. To build artificial information systems which reflect with high fidelity the perceived information systems.

3. To employ appropriate ontologies to rewrite these artificial information systems in the form of communication spaces which are publicly accepted by an interested community of agents – who are typically

    a. Peer scientists who are going to

        i. Check the proposed communication space with respect to similar portions of the reality those peer scientists observe independently of the first scientist; and

ii. Make use of the proposed communication space to guide their own perceptions of other portions of the reality; and

b. Engineers who are going to trust these communication spaces as representations of predictive models of behavior of certain portions of the reality, and use the communication spaces as communicated *information to command* (see Chapter 3 on the purposes of communication).

The trade of the scientists is, therefore, the construction and the exchange of information systems, among themselves and between themselves and engineers. The goal of science is the construction of artificial information systems – usually called *models* in the literature of Methodology of Science – capable of accurately predicting the behavior of carefully selected portions of the reality. A well selected portion of the reality, from the standpoint of scientific modeling, is a slice of the reality such that the interaction of the observer with reality, the observer's goals, intentions and plans, the perception of the observer about the other agents with whom it will interact, and the interaction procedures can all be deemed irrelevant.

Scientists are blissfully happy when they can detach themselves from their perceptions, and interact with other agents – e.g. other scientists and engineers – solely through communication spaces and specifically for collaborative problem solving. It is no surprise, therefore, that technological devices and conceptual tools for the construction and representation of communication spaces, facilitators for productive dialogues and ontological reasoning have been deemed so useful for scientific development.

Our radically heterodox view, however, is that science – and scientists – do not promote *knowledge*. Science is devoted to the development of a greater *understanding* of reality, dealing specifically and exclusively with *information*.

The development of science, however, must have been motivated by some very powerful necessity, lest our society would not have devoted so much energy along History for this path to the understanding of reality.

Up to this point, our book seems to be leading to the inevitability of isolation. Agents do not interact with each other directly, but through information systems, which create a mist of uncertainty between each other and isolate each agent from the world and from the other agents.

What could be the reason after all for people to build information systems, understand reality, and so on? Our proposition: these activities – scientific and technological development included – are intermediate resources for people to collaboratively reach each other, and ultimately reach the world itself. Information systems are therefore mediators to permit that people constructively and incrementally reach the world. When a person finally reaches the world, that person will *know* the world. Man will in this point finally become one with the world, with other men and with himself/herself. This is the point to be reached. This is the justification and the explanation for all the effort. Man detaches himself/herself from reality as a path to reach unity with the world that is manifested through reality, at which point any mediation is no longer needed. The ultimate measure of success of information systems shall be whether they reach the point in which they become unnecessary.

In slightly different words, the reason for an agent to sense information systems, to design, build and interact with reality and with other agents through artificial information systems, to collaborate in the construction of communication spaces, to solve problems collectively and cooperatively, is to find means to meet other agents and interact with them. Incidentally, interesting things happen on the way, such as the construction of conceptual tools and technological devices such as ontologies and computer systems to represent them as software applications, as well as the emergence of social behavior as the result of interaction and partial trust. The driving force, the wellspring of energy and the *raison d'etre* of all these processes, according to our view, is ultimately so that people can meet each other and interact with each other.

Why would people want to do that? Because people wish, ultimately, to really *know* each other. People want to know each other, they want to be known by each other, and therefore they want to share their knowledge. This is why they sense and build information systems and work collaboratively.

Information systems, and most remarkably artificial information systems implemented as digital devices, have opened new possibilities for communication among agents and relations with the world. This is new and wonderful, especially when it is explored as the means to reach a true communion of people and the world. In this case, these information systems lead to the acceptance of plurality and the enrichment of understanding and of interactions. The alternative to be avoided is to be enslaved by these artifacts, and therefore departing from unmediated contact and moving towards unified relations with everything – including oneself. Words such as *globalization* in certain sense denounce the choice for this path.

The illusion of life

Consider two brothers, who we are going to call Hansel and Christopher. Hansel is thirteen years old, Christopher is four years old.

There are strong reasons to believe that Hansel is going to be either a physicist or an engineer, as he has shown interest and talent to the physical sciences since very young. Christopher, on the other hand, seems very skilled in fine arts and music, and does not show much interest in mathematics or physics.

Their parents have taken Hansel and Christopher to the Disney World for the first time. They were both delighted by the experience, in quite different ways.

Hansel appreciated the technical quality of all the attractions in Disney World. The impersonations of all Disney characters looked real, the 3D movies really pushed the audience into the illusion of being into the action, and so on and so forth.

Christopher was simply marveled by the experience, and dived into the fantasy proposed by each attraction as if it were real. He interacted with each Disney character as if that character had jumped from the movie screen to reality – the characters were really there for Christopher, instead of people dressed up as cartoon characters; through the experience of 3D movies Christopher really lived the experience of participating in the featured stories – he *was* there, surrounded by his preferred characters and living with them the featured adventures.

What we have in every attraction in Disney World is the proposition of artificial information systems to the audience. As always happens, a proposed artificial information system can be perceived by an agent from different perspectives, depending on the context selected by the agent to appreciate the proposed information system. The context comprises the reality that is taken to be perceived by the agent, of which the proposed artificial information system is a part, as well as the perception of the agent about itself, based on which the agent understands its goals, purposes and capabilities.

Hansel was more interested in the physical reality that surrounded him, so that was the reality he perceived. He selected the appropriate information systems to perceive that portion of the reality, which included the artificial information systems proposed by the Disney World attractions. Hansel perceived some facets of those artificial information systems and added those facets to his previously existing framework of understanding of physics, based on which he built his understanding of the events that occurred around him.

Christopher, on the other hand, was more interested in the imaginary world proposed by the authors of the Disney World attractions. He selected different information systems to perceive a different portion of the reality, which also included the artificial information systems proposed by the Disney World attractions. Christopher perceived other facets of those artificial information systems and added those facets to his previously existing framework of understanding of the magical worlds proposed by the authors of the Disney attractions. Based on those perceptions, Christopher built his understanding of the portion of the reality that included the authors of the attractions, their imagination, their values as presented through their fictional works, as well as his own values presented to himself metaphorically through the characters involved in each specific attraction.

The attention of Hansel was focused on facets of the reality that were different from those that were the focus of interest of Christopher. Both children were relating to the reality through information systems, albeit following different paths to reach reality. Our point is that, in principle, both paths are equally good to understand reality. Both paths are also equally limited, in the sense that both Hansel and Christopher relate with the reality mediated by information systems.

Which path takes our children closer to knowledge sharing, instead of simply information exchange? Let us take Christopher, for example. To what extent is it possible for Christopher and the author of one of the Disney World attractions share knowledge?

That evidently depends on Christopher as well as the author of the Disney attraction. If that author prepares the script for an attraction assuming a "very professional" stance – in other words, detaching herself from the task of developing a Disney attraction – it is unlikely that her personality, beliefs, goals and purposes can be reached through the attraction. If on the other hand she "commits her guts" to the piece of art she is preparing, challenging herself every morning to prepare the masterpiece of her life that day, opening her soul to what she is building, then each artificial information system she prepares portrays the essence that defines herself – at least at that moment when she is preparing one specific attraction.

If Christopher, on his turn, opens his heart to the experience proposed by the author of the Disney attraction, and dives into the proposed experience as if it were real, then it *does* become real to him. The openness of both agents involved in this interaction gives room for a quasi-magical experience, through which they become momentarily one, sharing their values, emotions, purposes and goals.

This is our concept of *knowledge sharing*: a state that is reached by two or more agents, or by an agent and the reality, in which they become one. This state, as the example of the Disney World attraction clarifies, can be momentary, and does not require that all involved agents share their knowledge simultaneously.

As the example of the Disney World attraction also clarifies, knowledge sharing can make use of information flow and all other concepts discussed in the previous chapters to occur. Knowledge sharing, however, is radically different from the communitarian construction of a communication space.

In order to clarify this distinction, it is useful to set more clearly what we mean by *knowledge*. This is what we do in the next chapter.

Bibliographical notes

Our rather superficial analysis of the methodology of science, especially as to what it has become as the everyday work of professional scientists, is mainly based on reflections about our own routines. Deeper reflections, considering the historical development of the methodology of science, can be found elsewhere. The interested reader can be addressed for example to (Popper, 1969; Kuhn, 1996; Feyerabend, 1988), in order to appreciate some well founded, yet quite contrasting among each other views about the rationality and the utility of the scientific method.

The expression "the illusion of life" comes from a very beautiful book on cartoon animation, written by two outstanding animators from Disney studios. This book is about animation techniques, but it is also about the relation between reality and representation, a topic very much relevant to the discussion presented here. The interested reader is emphatically invited to consult this great book (Johnston and Thomas, 1995).

A vast literature can be found explicitly dealing with "knowledge sharing". This literature stems from two different academic traditions, connected respectively with Management Science and Computer Science. Some relatively recent contributions related to each of these traditions are, respectively, (Tiwana, 1999) and our own previous book (Correa da Silva and Agusti-Cullell, 2003). Both books contain a large bibliography, pointing to the major references related to each of these two academic traditions. We stress, however, that the common notion of "knowledge sharing" is akin to the notion of information exchange based on communication spaces, as presented in the previous chapters. The notion of knowledge sharing presented in this book – and more specifically in the present chapter of this book – differs very significantly from the standard use of this term. This differentiation is indeed one of the main reasons we wrote the present book.

Intermission

The machines arrived this morning.

The lateral wall is completely gone.

Yesterday evening the architect who is responsible for the whole project came here to speak with Dr. Agnelotti.

She is a very young lady. Her eyes are incredibly green, and she has long blond hair.

When she speaks, she sounds as if she could not be that young. I am really convinced she truly knows what she is doing.

Would she sound so competent and convincing if she were not so beautiful?

Who do we trust, and why?

She came to explain about their plan to preserve the fig tree. They thought it was too beautiful to be cut down, and it could suffer too much if they tried to move it to another place.

They had made a slight change in the design of the road, so that the fig tree could stay as it was.

And we did not even have to ask.

I started to believe that the director plan was in good hands.

We woke up a little earlier than usual this morning. Everything had already been set, but still we thought we should better have a last check before the work started.

The tomatoes had been all picked. The chicken had been given to Mrs. Agnelotti's sister. We had taken a lot of parsley, basil and chilies. As much as we could. It was not the time of plums or figs.

We had cleaned the floor and organized all tools and objects in cases.

The back garden was clean.

We had breakfast under the fig tree, as usual. After breakfast, we moved the garden table to the corner close to the kitchen door.

The door bell ringed. This time it was the engineer who would look after the work today.

He was very kind. He asked whether they could start working on the lateral wall. The plan was to have it down before lunch time.

Mrs. Agnelotti offered him a cup of coffee. He accepted it. Then he went out and we could hear him explaining the general plan to his colleagues.

We did not talk much that morning. We worked and organized things. There was a lot to do. We were busy. We did not feel like talking.

We did not feel sad, though. At least I did not. My impression was that serenity had fallen on us like a blanket.

This is an important day. A moment to be remembered. As so many moments in that back garden.

The lateral wall is gone. And it is not even lunch time yet.

I take some pictures.

I stay close to the fig tree. I feel like standing close to an old friend.

Mrs. Agnelotti suggests that we go to have lunch. Today we have all been too busy the whole morning and we could not prepare lunch. We did not feel like preparing lunch.

We were going to a restaurant.

As always, Mrs. Agnelotti had managed to transform a difficult moment into a festive situation.

We were going to have a good time in a very nice restaurant. She had already suggested the restaurant, the food and the wine.

We organize our things and leave.

After lunch Dr. Agnelotti and I start to put things together. The result of our work.

BackGarden version 1.0.

We have taken many pictures. We have recorded some sounds. We have recorded some of our conversations. I insisted that it would be nice to have Dr. Agnelotti and Mrs. Agnelotti recorded telling some of their stories. It would be nice for Giovanni to hear his grandparents telling those stories.

Above all, we have made a lot of notes. We have written a lot. We have drawn diagrams. We have annotated a lot.

We have talked.

It is all written down. We have not missed a word. A sound. A breath.

The back garden is saved.

Our pictures can be helpful. The sounds are a nice document, too. The voices of Dr. Agnelotti and Mrs. Agnelotti are touching. Of course we only recorded a couple of stories. I believe Giovanni will appreciate them, though.

Our notes can be interesting too. They have become rather voluminous. They are relatively well organized, nevertheless. There is a sense of order in them. They form a grand picture of the garden.

There are also the diagrams and footnotes and methodological explanations.

I like to believe that there is an intrinsic value in them, too.

Our notes are nice. It becomes evident that there is a lot of work in them.

They are scattered, though. They talk about a few points and details.

The back garden is saved, though. To its entirety.

It took me quite a while to realize what we have been doing.

Actually, it was only this morning that I finally got it.

I was standing by the fig tree when it came to me. I believe it was the fig tree who decided to confide one of its secrets to me.

I suspect I was the last one to understand. I am very much convinced that Dr. Agnelotti knew that. Perhaps he did not know when we started this work. At some point he realized the whole thing.

I suspect that Mrs. Agnelotti was the first one to understand. She waited patiently until we all reached the same conclusion.

Certain things are not to be told.

Certain things are only to be unfolded.

When the time comes.

Not earlier. Not later. At the exact time.

Giovanni probably knows everything. He has the wisdom of a six months old person.

We forget things on the way.

We sometimes decide to delude ourselves.

We start to believe in illusions.

Much worse, we tell those illusions to the ones who still can see the truth.

We corrupt them.

We move them from the way.

So that we can all move back to the way later on.

Strange world, this one.

Strange world, this one we build day in day out.

We have taken notes and pictures and all. We have been writing down everything about the back garden.

We have not written about the back garden in our notes.

The images of the back garden are not printed in our pictures.

Our learning and understanding about human nature is not in our diagrams.

Our messages sent to our friends in Botswana and Bangalore are not in our notebooks, or in our computer.

The back garden is in us.

The back garden is us.

We needed these notes and pictures and conversations. Without them, we would not have managed to become one with the back garden.

Other people might be able to do that without a laptop and a digital camera. Perhaps our friend in Bangalore could do it.

Not us. We needed exactly this experience.

Writing notes was essential for us to become one with the back garden. Yet, now that we are one with the back garden, the notes – as interesting and entertaining as they can be – are not fundamental to convey the back garden to Giovanni, or to any one.

The back garden is not in our notes. It is in us. As we are in the back garden.

The Agnelotti back garden cannot be destroyed. It is truly indestructible.

What is the back garden?

The back garden itself is another experience. The back garden is a perceptible experience to us. Through it, we have found something else.

We bow and thank to the back garden. It has shown us important things. About ourselves. About being human. About being alive.

Others may not need a back garden to understand what we have started to see through this back garden. Perhaps the farmer in Botswana needs his herd of cows to meet us at this juncture in our paths through life. Perhaps the grocer in Bangalore needs his wife and children to meet us at this same juncture.

Or perhaps the grocer in Bangalore prefers to meet us a little steps – or miles – ahead on our ways.

The back garden has no start, and no end. The back garden is everywhere. It is beyond space. The back garden has always been here, and will always be. It is beyond time.

We have unveiled the timeless and spaceless back garden in which we live. Thanks to this particular back garden, full of parsley and basil and plums and chicken and gifts hanging from the fig tree and gates broken just to reveal love, sweetness and respect.

The Agnelotti back garden has been generous to us. He has taught us a lot.

There are quite a few things now to tell Giovanni-at-the-age-of-five.

Funny how we need something as rough as a director plan from the municipality of San Isidro to start looking around.

Funny how we need to look around to start noticing the details.

Funny how we need to pay attention to the details to start seeing what goes beyond the details.

We have been through a very long way. We have traveled a lot in the past few days.

We have chosen a rather tortuous path.

We are here, anyway.

At least we believe we have reached some important landmark.

A resting post, at least.

Before we continue our journey.

There is still a lot to learn about the back garden.

We thought we would have things to tell to Giovanni. Giovanni is six months old. He has not forgotten much. Not many people have had the chance to annoy him with illusions and fake images.

We probably could have taken a much shorter path.

We could simply have asked him.

He could have told us.

He could have explained us about the back garden.

We have written down our annotations nevertheless.

Gigino here, Cicilla there, the chicken, the plums, the tomatoes.

The fig tree.

I believe he will appreciate that.

A last fine detail about the director plan.

The new road does not go through the ice cream shop.

We can still go there this afternoon.

Tomorrow morning I leave back home.

It has been a fine stay, this one here in San Isidro.

I have learned quite a few things.

Un pour tous, tous pour un, the three musketeers would say.

This motto was about friendship.

One for all, all for one.

One for everything, everything for one.

One is everything, everything is One.

So long, San Isidro.

Good morning, San Isidro.

Chapter 9

Knowledge

As I write this book, following a tale told in an ancient almost illegible chronicle, I realize only now that I have filled page after page and am still at the very beginning.

Italo Calvino

What is knowledge, then, after all? It is not what defines a person, but it is inseparable from a person. Knowledge cannot be perceived as an information system to be encoded and transmitted through a communication channel or stored as a collection of computational data structures, and therefore it does not make sense to talk about knowledge communication, knowledge flow, or knowledge base. What does make sense is to talk about knowledge sharing, as discussed in the previous chapter, as the creative act through which a community of agents work together to transform their individual selves, accepting as a foundational presupposition that each individual self cannot be separated from the community and from the surrounding reality.

Despite the fact that we nowadays very often replace knowledge by information, they are very different. For practical purposes, we can replace facets of the reality such as *water*, *air*, the *Earth* and the *Sun* by their scientific, i.e. information-mediated descriptions, e.g. water by its abstract description in terms of its chemical components, H_2O. One can drink water, or wash oneself with water, but not with the abstract formulation of water as H_2O.

Knowing something means to become that something that is known. An information system cannot be a substitute of the reality. At its very best, an information system reflects some dimensions of the reality. Information is an inseparable component of the reality and therefore of knowledge, but a component of a system cannot replace the whole system.

Information systems are perceived, recorded and built for specific purposes. Man relates to the reality through information systems in order to satisfy purposes, reach goals and execute plans. Knowledge, contrastingly, is not instrumental. Knowledge comes together with a sense of freedom from necessities and purposes to be satisfied.

The understanding of science as a disciplined methodology to build information systems is in accordance with the view that science works hand in hand with technology, and more often than not is guided by technology. The development of modern science – and particularly the institutionalized, "professional" science – comes in response to a technological attitude towards the reality. Science provides the required information to enable technology to transform the reality, so that the purposes, goals and necessities of people can be satisfied. The power to transform the reality through technology, in turn, reassures science as an effective means to approach the reality, and provides to it a pragmatic sense of truth.

Information is a coin with two faces. One face relates to concepts and abstractions, and we call it Science. The other face relates to the embodiment of concepts in devices, social practices and norms. We call this other face Technology.

Knowledge implies purposeless communion with reality and with other agents. Through knowledge, the perception of the boundary that separates each agent from the others and from the reality reveals itself as an illusion, an artifice that is useful for the construction of abstractions which in turn enable the construction of information systems.

Knowledge is what connects all agents with each other and with the reality. Not only with the manifestations of the reality – *aka* information systems – but with reality itself. Each individual agent, as well as reality, are ever changing, therefore knowledge is inherently dynamic.

Knowledge manifests itself in the connections among agents and reality. It can be found, therefore, in the relationships and bindings among agents, and between agents and reality. Differently from information and information systems, knowledge defies formal representation, due to its fluidity and ever changing nature.

Our conceptualization of knowledge characterizes it as fundamentally distributed. There is no "my knowledge", "your knowledge", or "that knowledge over there", bottled in a container or transcribed as a collection of data structures. There is simply Knowledge, as the glue that puts together all agents and connects them with reality.

Knowledge is the supporting material and the driving force that springs agents to connect with each other and with reality. Agents then employ the resources they have at hand to construct relationships, as intelligible images of the connections they fundamentally look for. The resulting process is the perception of information systems and the construction of webs of relationships based on information systems, artificial information systems, communication spaces, and so on.

A most visible side effect of the constructed relationships is our present day society, in which Technology and its sister Science play such central role, supported by information systems which have recently moved from abstract concepts to the status of very concrete devices and appliances – such as computers, the Internet and the ubiquitous World Wide Web. These side effects, as useful as they can be, should nevertheless be taken as what they are, i.e. rather peculiar incidental steps that have been taken so that the fundamental thirst of us people could hopefully be appeased. The thirst for connections. The thirst for Knowledge. The thirst for Happiness.

Bibliographical notes

The definition of knowledge has kept busy many philosophers for a long time. Our way to contrast information and knowledge has been influenced by, among others, (Borgmann, 1999). Our notion of knowledge, in turn, has been influenced by many sources, of which we highlight the work of R. Panikkar, e.g. (Panikkar, 1975) and (Panikkar, 1980), and the sacred books especially of the Hindu and Christian traditions.

Chapter 10

Conclusion

"Is there nothing salvageable, then?"
"Maybe. But not here."
"Who? Where?"
"The knights of the Holy Grail."
And where are they?"
"In the forests of Scotland."
"Have you seen them?"
"No!"
"Then how d'you know about them?"
"I know."

Italo Calvino

As we approached the end of this book, we have realized that it has turned to be rather different from what we had planned and from what we expected. At first, we wished to contribute to the technical literature on the design and implementation of software applications most commonly called either *knowledge based systems* or *knowledge management systems*. This would be, after all, a natural follow-up of our previous work. Our dissatisfaction with the terminology used for these systems, and later with our own shallowness in previous analyses of these systems – their motivations, design drivers, and consequences – turned this book to a more reflexive and nearly philosophical mood.

We started writing a contribution to the technology of a specialized sort of software development, went through a discussion about the nature of the software we wished to see well built, and ended up in considerations about the reasons why someone would want to build such software. At this point, we realized that these reasons were far more encompassing than just reasons to justify the construction of this or that piece of software.

We discussed for quite a few years, and wrote about three times the number of pages you see in this book – which became abandoned drafts – before we decided to work on the text you have now in your hands. One of the reasons – probably the main reason – we hesitated so much in facing the issues we did in the previous chapters was that we did not feel prepared to face such challenging issues. We still do not feel prepared at all, to be honest. We felt, however, that if we did not strike directly on these issues we would feel so deeply dissatisfied that we would most likely not be able to write anything else, at least not as wholeheartedly as we did in this book. It had taken us too much effort to transform the haunting ghosts that inquired us in the past – as mentioned in the Preface – into loving companions, and we would not risk seeing our companions turned into spooky creatures again.

We had two challenges in the preparation of this book: writing with acceptable competence about these issues, and finding an adequate format to present these ideas. We did not want to give up the clarity and precision of scientific writing, yet we wanted to write a book that could be read by anyone interested in these issues, even those who were not trained philosophers or PhDs in Computer Science. This second challenge led us to use the rather uncommon style we did, in which we avoided technical and mathematical language, interspersed the whole text with concrete examples and short tales, and prohibited ourselves from quoting other authors along the chapters. We did our best to identify our sources of ideas and inspiration in the bibliographical notes at the end of each chapter, which we also hope can be useful for the interested reader to study further specific topics mentioned in this book.

After many years of discussion and many pages to present our ideas, however, we have found a short text that synthesizes much of our conclusions. We granted ourselves in the conclusion of this book an exception to the "no quotes" rule, to bring to you this short text. We really do not know what effect this text can have in you, dear reader, but we confess that it was reassuring to us to find in an ancient poem ideas with which ours resonate so well. Better yet, the author of this poem renames our notion of knowledge, using *Love* as a better name. We did not change the name in the previous chapters, lest our book could end up classified in the wrong bookshelves in libraries and bookstores.

We thus conclude this book by quoting a poem by Saint Paul the Apostle, found in one of his Letters to the Corinthians, hence written way before the advent of the Information Society, Computer Science, Artificial Intelligence, Knowledge-based Systems or Business Intelligence.

1. *Though I speak with the tongues of men and of angels, and have not love, I am become as sounding brass, or a tinkling cymbal.*

2. *And though I have the gift of prophecy, and understand all mysteries, and all science; and though I have all faith, so that I could remove mountains, and have not love, I am nothing.*

3. *And though I bestow all my goods to feed the poor, and though I give my body to be burned, and have not love, it profiteth me nothing.*

4. *Love suffereth long, and is kind; love envieth not; love vaunteth not itself, is not puffed up,*

5. *Doth not behave itself unseemly, seeketh not her own, is not easily provoked, thinketh no evil;*

6. *Rejoiceth not in iniquity, but rejoiceth in the truth;*

7. *Beareth all things, believeth all things, hopeth all things, endureth all things.*

8. *Love never faileth: but whether there be prophecies, they shall fail; whether there be tongues, they shall cease; whether there be science, it shall vanish away.*

9. *For we know in part, and we prophesy in part.*

10. *But when that which is perfect is come, then that which is in part shall be done away.*

11. *When I was a child, I spake as a child, I understood as a child, I thought as a child: but when I became a man, I put away childish things.*

12. *For now we see through a glass, darkly; but then face to face: now I know in part; but then shall I know even as also I am known.*

13. *And now abideth faith, hope, love, these three; but the greatest of these is love.*

Epilogue

We are approaching midnight.

It has been a long day.

Tomorrow I will have to wake up very early again. I will have more lectures to present.

Winter is approaching. It is still dark when I leave to the University.

There should be a law against this.

This morning there was a rather annoying student in the class. He laughed and talked all the time. I considered his attitude a little impolite, at the verge of confrontational.

He was tall like a giraffe. His furry sweater made him look like a marmot. His voice sounded like a macaw. He laughed like a hyena.

We were discussing infinite loops and program termination.

At some point, for a reason I cannot remember, we touched on the meaning of the word *success*. From that we moved to the value of culture. And then to values in general.

Funny how a discussion about the organization of procedures within the code of a program can lead to this and then to that and to that and suddenly we are talking about values and culture.

I ventured to ask my students why they were studying so hard at the University anyway.

The inconvenient student replied with a clear and loud voice: "I am here because I want to be rich", and burst into laughter.

Some of his colleagues – the majority, to be honest – laughed with him. We could hear some "Yeahs!" here and there.

To my relief, a few students seemed a little disgusted with that.

No one dared to challenge the giraffe. His macaw voice resounded in the room. His hyena laughter could be heard together with the laughter of his pack.

I kept silent for a while.

It took me a while until I could decide about any sort of reaction.

Then I sighed.

I felt like moving back to the discussion about how to avoid infinite loops.

The voice did not let me, though.

There is a funny voice that comes from some part of my body sometimes. I am sure you know about what I am talking.

Sometimes I think it comes from my stomach. Other times I suspect it comes from my head. I like to believe it comes from my heart. From what it tells me sometimes, though, I would rather have it coming from a less noble part of my body. Perhaps my toes or my knees.

The voice told me I could not let it go.

The voice told me I had to express my position.

I looked again at the giraffe. I sighed again.

"That is fine, as long as you define what you mean by 'rich'." This was the best I could find to tell the giraffe and all the students.

"By 'rich', obviously I mean full of money" the macaw voice stated again. The hyenas laughed even more loudly.

Some of the students who looked disgusted at first now laughed together with the hyenas. The very few who still did not laugh looked at me, as if I were a knocked down boxer. The counting was on. I had to stand up, or it would be a knock out.

I had to ask questions for which I did not have all the answers.

A few minutes ago I was thinking it was going to be an easy lecture. A few mathematical properties, a few programming constructs, and that would be it.

Now I had to hold a giraffe and a pack of hyenas.

I had to act quickly.

I recalled the story about a Karate master. A true story.

One day a new student came to his dojo. He was young and arrogant. He was also very big and strong.

The Karate master was short and slim.

He let the student stay in his class.

The student beat all other students.

At the end of the class, the arrogant new student approached the master. He said “one day I am going to have a black belt just like yours”. The Karate master did not say a word. He untied his own black belt and offered it to the new student. The student did not accept the black belt. He left the dojo ashamed and never returned again.

A black belt is not worth for what it is. A black belt is worth for what it represents. You may desire what it represents, you may have what it represents, and you do not need to have the black belt itself. You may have the black belt and not have what it represents, and then you have nothing.

I asked what money meant for each of them.

The voice inside my belly silenced. I could finish the lecture now.

I finished my lecture and left the room hastily.

I had had enough for one day.

I have however been thinking the whole day about the meaning of money. Or the many meanings of it.

I must remember to thank the giraffe for bringing this issue about. I feel like sincerely thanking him for that. I also consider it can be healthy for him to realize that his participation in the lecture has been given importance, in a constructive way. He was not simply a nuisance. It can help him consider his own attitudes and participation in the lecture with a little more responsibility.

I must remember not to call him "giraffe", though.

It is funny to consider how money operates.

I prepare a lecture.

The baker around the corner bakes some rolls of bread.

The tailor makes a shirt.

A writer writes a novel.

My lecture is useless without students.

Bread is useless if no one eats it.

A shirt is useless if no one wears it.

A book is useless if no one reads it.

I prepare and present lectures because this is what I know how to do. I do not know how to make a shirt.

The students make my lectures useful. They make them worth something. I should thank them. I should pay them back by their generosity in making my lectures – and therefore myself – useful in society.

Yet, I get paid for my lectures.

Then I get my money and buy a new shirt.

Should not the tailor be grateful to me, because I am making his work – and therefore himself – useful? Should he not be the one to pay me by wearing the shirt he has made?

Why does it sound so strange? Why is it not that I go to the tailor and choose my shirt, and if I choose the most beautiful one he pays me the most, and then with that money I leave the tailor workshop all happy because now I can pay back the kindness of my students, who come to my lectures?

We have however turned the arrows the opposite direction.

I get paid by my students for the lectures I give. With that money I go to the tailor and pay him for the shirt.

A very strange equivalence relation is built between my lectures and shirts.

I stay with my students for about three and a half hours and that is worth a shirt. I can teach my first year students to write a procedure that, given *N*, calculates the *N* first prime numbers in three and a half hours. Including a few variations that improve the performance of the code a little bit.

The *N* first prime numbers are worth a shirt. They are also worth a novel that took five years to be written, and later earned the Pulitzer Prize.

These equivalences are stated using money. Money is the language that allows the expression of the equivalence between a couple of lectures, a shirt and a novel.

Money is therefore a language. Actually, money is not a complete language. Money is a word. A single word that expresses a simple convention, through which people state how much is worth their work, according to themselves.

In a funny reverse way, as discussed above.

Considering this way, it is strange to think that some people spend so much time and energy accumulating money. It could be more interesting to build a collection of dictionaries. One would be collecting words either way.

Apart from the consideration that remuneration seems to be reversed in our economic system, money seems therefore to be important as a means for people to inform each other how much they consider that certain objects or activities are worth.

It is a standard terminology that people use to say "what you have built is good" or "what you do is important".

I do not need money to say anything like that to my sister. I simply go and tell her "what you are doing is important" and "what you are building is good". She knows what I mean by that.

Especially if I say things like that to her during lunch on a Sunday. Especially when we are one table apart.

If however I try to say the same thing to a businessman in China, things can be a little more complicated.

We need a convention. We must find a way to express ourselves, so that we can understand each other.

There is no way I can express to the businessman in China the same wealth of messages I convey to my sister when I tell her "what you are doing is important" during lunch.

We must accede to being less expressive. It is this or no communication at all.

We prefer the poorer communication.

We use, for example, money to communicate.

I express to the Chinese entrepreneur how much I think his product is worth, and he understands my message. In order for the message to get through and be understood unambiguously, it has to be short and relatively crude. No subtleties allowed.

We accept to say less, and with that we can say things to more people.

Mass communication. Money is a means for mass communication.

People like money because its message can be quite precise.

The problem is, some people like money a little too much.

Take a horse. Feed it with sugar all the time. Only sugar. No vitamins, no proteins, nothing else but sugar.

The horse will feel satiated. It will not feel hungry. Yet, the horse will have serious nutrition problems after a short while.

It will crave for food all the time. It will need vitamins and certain minerals. It will eat more and more sugar. It will not work.

Take a person. Feed her with comfort. Feed her with electronic gadgets. Feed her with mansions.

Feed her with money.

Is she not satisfied after a while?

Maybe she is being fed with the wrong nutrients.

Maybe you are feeding a horse with sugar.

What could be an interpretation for *savings*?

Let us consider for a while that upside down economy we referred to before, in which the tailor pays some money to the person who uses his shirt, as an expression of gratitude for making his work useful.

If I suddenly have a lot of money in this economic system, it means I am consuming more than I am producing. I should work harder, to make myself more useful. If I worked harder, I would have the opportunity to be more useful. I would grant more people the opportunity to use the result of my actions. I would have more people to thank.

I would have the opportunity to give my money to more people.

Consider now our "actual" economy.

It is the opposite of the proposed upside down economy.

Therefore, if I accumulate money, it should mean that I have been making myself useful more than some other people – those who gave me their money.

Accumulated money should indicate greater utility.

Look around. Is this what you see?

Where is the catch?

Could it be the case that money has detached itself from its original meaning – a message indicating an expression of gratitude, either in the proposed upside down economy or in the "actual" economy?

Could it be the case that, by detaching itself from the web of inter-relations that granted its existence, money has become a value in itself?

Perhaps money has changed its meaning through History?

Could it be that we find problems and distortions in our society whenever some part of it claims *independence* from all other parts, thus refusing its noble position in a web of inter-relations, to reclaim to itself undeservedly a superior position?

Is it not what occurred with this interesting and highly successful technological device called *money*?

Why is it that we admire and even envy those very wealthy people who appear in the lists of publications like *Fortune*, but we do not quite feel like having any of them as our son in law?

What if money became simply the means to indicate who is making himself more useful to whom?

What about computers, the Internet and the World Wide Web?

What about 3-D virtual worlds and the alternate digital reality they propose to people?

What if people kept these devices in perspective, not allowing the devices to distract them from the point of meeting each other, connecting to each other, communicating with each other?

What if people considered technological devices, and Technology itself, as appropriate means to reach greater and more ambitious goals?

Goals like those about which we sometimes think in those quiet summer evenings, when something reminds us that we are probably not going to be here less than sixty years from today, and that we should better start thinking whether our life has been that of producers or consumers?

A producer grows old and, all things going well, one day is capable of looking back in life and saying, lying on his bed, "boy I feel good about my life. I have made the difference to a lot of people. I shall be remembered by the results of my actions. I have planted seeds that I now can see are growing strong. I see people living happy and I have contributed to their happiness."

A consumer grows old and, all things going well, one day looks back in life and thinks gravely, lying on his bed, "oh dear! It is too late now. I no longer have the strength to do anything else. I look around and have not made any difference to anyone. I shall not be remembered. There is not much left for me to do. I should better leave stage quietly now."

What if people started deciding to be producers earlier in life?

Would we consider the importance of accumulating goods differently?

Would we consider the importance of looking beautiful in virtual worlds differently?

I think again about my student, the giraffe who talks like a macaw and laughs like a hyena.

Do my erratic thoughts and considerations make any difference to him?

What would it take for me to lean to the producers' side as a lecturer at the university?

Are universities good institutions? Are they useful at all? Do they help building producers?

Are we university lecturers only building consumers?

What is the role of technology in our lives?

What is technology?

What has science to do with this?

What has the university to do with this?

The end of another day. My hair is very grey now. It seems to be growing greyer every day.

It is quite late.

Again.

I should better get some sleep.

Again.

Tomorrow I have to wake up very early.

Again.

I shall sleep with my questions.

Again.

They shall wake me up tomorrow.

Again.

Caminante, no hay camino,
Se hace camino al andar.

Antonio Machado

Valeu a pena? Tudo vale a pena
Se a alma não é pequena.

Fernando Pessoa

Try again. Fail again. Fail better.

Samuel Becket

Bibliography

Bicchieri, C. *Game Theory: Nash Equilibrium.* In Floridi, L. (ed.) *The Blackwell Guide to the Philosophy of Computing and Information.* UK. 2004.

Borgmann, A. *Holding on to Reality: the Nature of Information at the Turn of the Millennium.* The University of Chicago Press. USA. 1999.

Colombo, E. *Knowledge Artifacts from an Artificial Intelligence Perspective.* PhD thesis, University of Milano-Bicocca. Italy. 2005.

Correa da Silva, F. S. and Agusti-Cullell, J. *Knowledge Coordination.* Wiley. 2003.

Dretske, F. *Knowledge and the Flow of Information.* MIT Press. USA. 1981.

Dutta, P. K. *Strategies and Games: Theory and Practice.* MIT Press. USA. 1999.

Fagin, R., Halpern, J. Y., Moses, Y. and Vardi, M. Y. *Reasoning about Knowledge.* MIT Press. USA. 1995.

Feyerabend, P. K. *Farewell to Reason.* Verso. USA. 1988.

Gruber, T. R. *A Translation Approach to Portable Ontologies.* Knowledge Acquisition, v. 5(2): 199-220. 1993.

Gruber, T. R. *Ontology of Folksonomy: A Mash-up of Apples and Oranges.* Invited paper / keynote presentation. First online conference on Metadata and Semantics Research. Online version can be found at *http://tomgruber.org*. 2005

Hayles, N. K. *How We Became Posthuman: Virtual Bodies in Cybernetics, Literature, and Information.* The University of Chicago Press. USA. 1999.

Kant, I. *An Answer to the Question: What is Enlightenment?* Original German publication in Berlin Monthly. Germany. 1784.

Kuhn, T. S. *The Structure of Scientific Revolutions*. University of Chicago Press (3rd revised edition). USA. 1996.

Mitchell, J. C. *Foundations for Programming Languages*. MIT Press. USA. 1996.

Moravec, H. *Mind Children: The Future of Robot and Human Intelligence.* Harvard University Press. USA. 1988.

Newell, A. and Simon, H. A. *Computer Science as Empirical Inquiry: Symbols and Search.* Communications of the ACM, 19 (3). Pgs. 113-126. ACM Press. USA. 1976.

Norman, D. *The Design of Everyday Things.* Basic Books (2nd edition). USA. 2002.

Panikkar, R. *Singularity and Individuality, the Double Principle of Individuation.* Revue Internationale de Philosophie, 111-112 (1-2). Pgs. 141-166. Belgium. 1975.

Panikkar, R. *Words and Terms.* In Olivetti, M. M. (ed.) *Volume in Honor of Professor Enrico Castelli.* Italy. 1980.

Popper, K. R. *Conjectures and Refutations: The Growth of Scientific Knowledge.* Routledge (3rd revised edition). USA. 1969.

Sharp, H., Rogers, Y. and Preece, J. *Interaction Design: Beyond Human-computer Interaction.* John Wiley and Sons (2nd revised edition). USA. 2007.

Solomon, R. C. and Flores, F. *Building Trust: in Business, Politics, Relationships, and Life.* Oxford University Press. USA. 2003.

Stewart. I. *Does God Play Dice? The New Mathematics of Chaos.* Blackwell (2nd revised edition). USA. 2002.

Thomas, F. and Johnston, O. *The Illusion of Life: Disney Animation.* Disney Editions (2nd revised edition). 1995.

Tiwana, A. *The Knowledge Management Toolkit: Orchestrating IT, Strategy, and Knowledge Management Platforms.* Prentice-Hall. USA. 1999.

Wenger, E. *Communities of Practice: Learning, Meaning and Identity.* Cambridge University Press. USA. 1998.

Winograd, T. *The Design of Interaction.* In Denning, P. and Metcalfe, R. (eds.), *Beyond Calculation, The Next 50 Years of Computing.* Springer-Verlag. USA. 1997.

Winograd, T. and Flores. F. *Understanding Computers and Cognition: A New Foundation for Design*. USA. 1986.

Index

www.ingramcontent.com/pod-product-compliance
Ingram Content Group UK Ltd.
Pitfield, Milton Keynes, MK11 3LW, UK
UKHW050946290926
15470UKWH00005B/1659

9 780444 529350